A Wood of Our Own

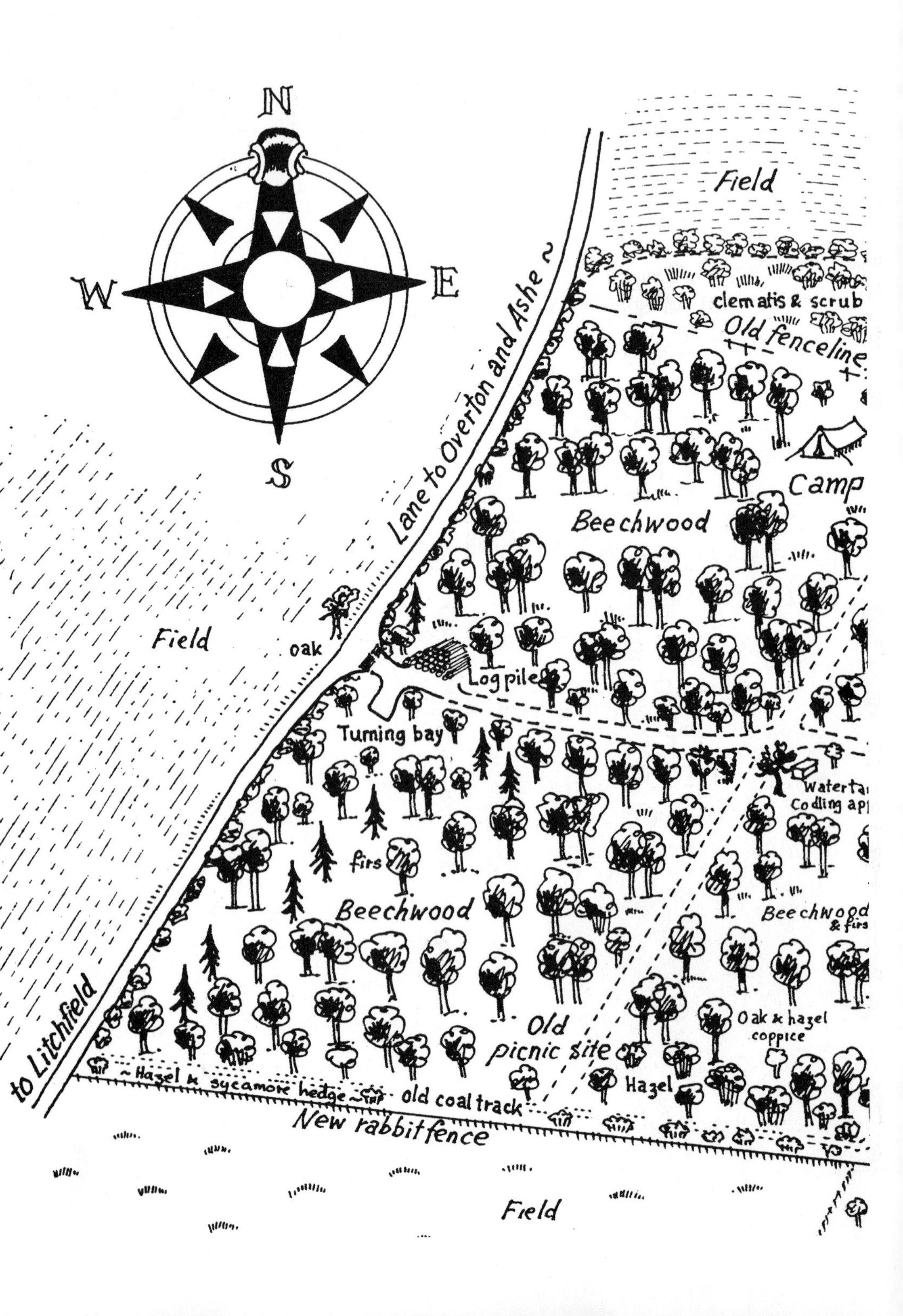
N
W
E
S
Field
Lane to Overton and Ashe
clematis & scrub
Old fenceline
Camp
Beechwood
Field
oak
Log pile
Turning bay
firs
Beechwood
Beechwood & firs
Old picnic site
Oak & hazel coppice
Hazel
to Litchfield
Hazel & sycamore hedge
old coal track
New rabbit fence
Field

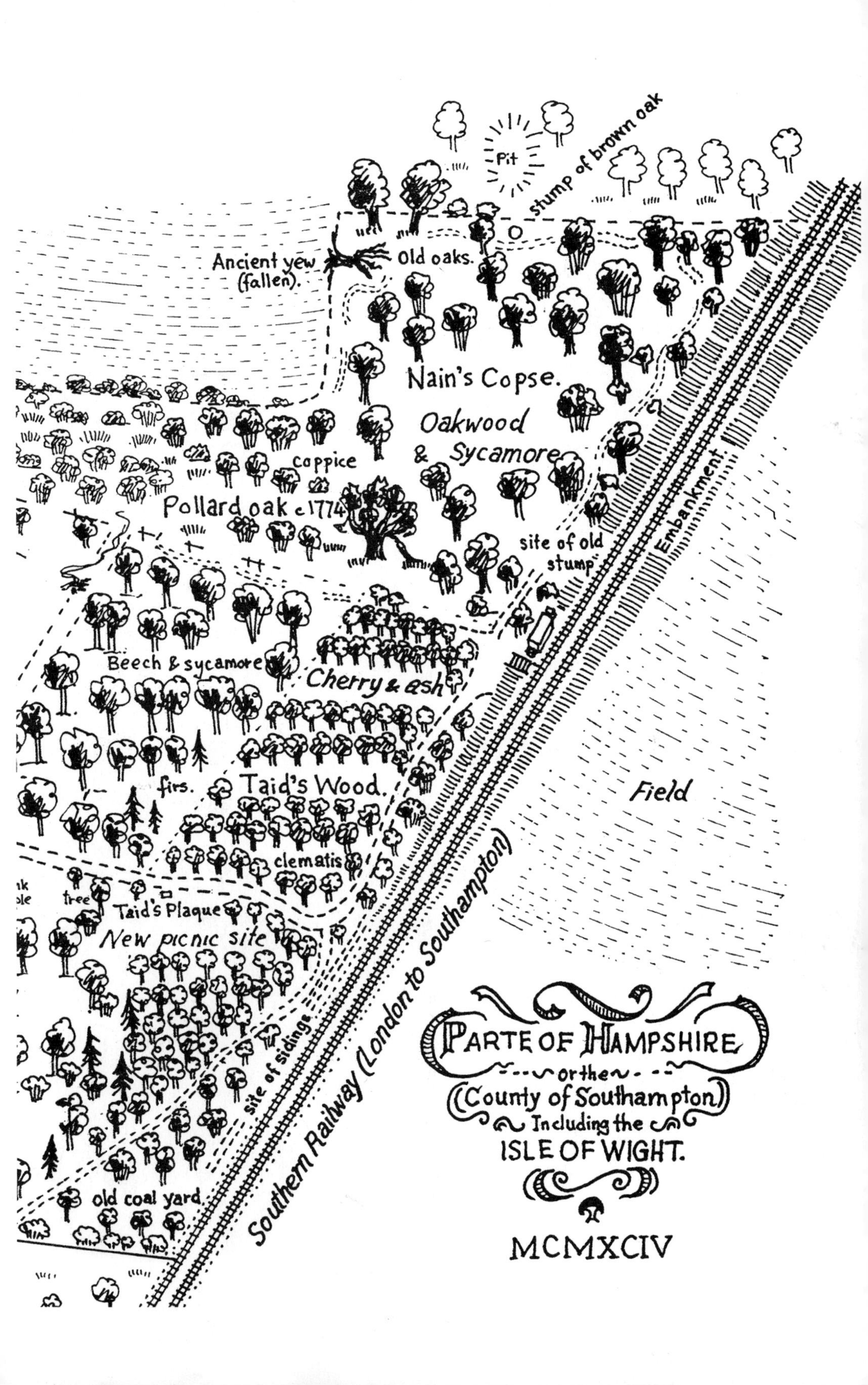
Pit
stump of brown oak
Ancient yew (fallen).
Old oaks.
Nain's Copse.
Oakwood & Sycamore
coppice
Pollard oak c1774
site of old stump
Embankment
Beech & sycamore
Cherry & ash
firs.
Taid's Wood.
Field
clematis
tree
Taid's Plaque
New picnic site
site of sidings
Southern Railway (London to Southampton)
old coal yard
PARTE OF HAMPSHIRE
or the
(County of Southampton)
Including the
ISLE OF WIGHT.
MCMXCIV

A Wood of Our Own

JULIAN EVANS

With illustrations by JOHN WHITE *and* STEPHEN EVANS

FOREWORD BY ALAN TITCHMARSH

Oxford New York Tokyo

OXFORD UNIVERSITY PRESS

1995

Oxford University Press, Walton Street, Oxford OX2 6DP

Oxford New York
Athens Auckland Bangkok Bombay
Calcutta Cape Town Dar es Salaam Delhi
Florence Hong Kong Istanbul Karachi
Kuala Lumpur Madras Madrid Melbourne
Mexico City Nairobi Paris Singapore
Taipei Tokyo Toronto

and associated companies in
Berlin Ibadan

Oxford is a trade mark of Oxford University Press

Published in the United States
by Oxford University Press Inc., New York

A catalogue record for this book is available from the British Library

Library of Congress Cataloging in Publication Data
(Data available)

ISBN 0 19 854951 2

Typeset by EXPO Holdings, Malaysia

Printed in Great Britain by
Bookcraft (Bath) Ltd
Midsomer Norton, Avon.

To
my wife,
my mother, and
my mother-in-law

Foreword

ALAN TITCHMARSH

Countryside without woodland would be a poor place—bare pasture open to all winds with nothing in the way of shelter for wildlife. There would be no bluebells, no badgers, and no birdsong.

But plant a wood and the landscape is transformed. Birds nest among the treetops and the scrub below. Badgers and foxes, deer and dormice find a home, and we have the thrill of walking below towering oaks and beech trees, among groves of wood anemones and bluebells, to discover rarer delights like wild orchids and flamboyant fungi.

My love for woodland began when I was a boy—regularly taken for walks in Middleton Woods on the banks of the River Wharfe in Ilkley, Yorkshire. It's a love that has grown and which, a few years ago, reached its zenith when I was able to buy and plant up a few acres of pasture to make my own broadleaved woodland, similar in size to the one Julian Evans describes and which I visited a few years ago.

He, too, knows the thrill of planting a 6-inch-high sapling and watching it mature with amazing speed into a young tree. In just two years some of my oaks and wild cherries, the ash saplings, and the field maples are already over six feet high, thanks to tree shelters and weather which has made sure the roots have not gone short of water.

Julian has owned his wood for ten years, and as the Forestry Commission's expert on growing broadleaved trees he knows how to look after it. You see, you can't just plant a tree and forget it. Woodland, even though it may be 'wild', needs a bit of looking after. The work is nothing like as intensive as gardening, but it is still important if the woodland environment is to work. Sadly, many woodlands are left completely alone by their owners who are uncertain what to do.

This story tells of what Julian did and touches on all facets of woodland life: the fun as well as the forestry. It will interest farmers who are wondering whether to put some set-aside down to trees. The present resurgence in planting farmland, mostly with native species like ash, oak, and wild cherry, is tremendously encouraging. It will help to restore our intensively worked countryside and to put back the hedges, spinneys, copses, and woodlands that are part of the British countryside we dream about. But the countryside captured in Constable's paintings was a working countryside: the story Julian tells is of a working woodland.

In a world of impatience, it is good to plant something that can be enjoyed by our children and grandchildren. But don't imagine that you will never enjoy it yourself—you'll be surprised how quickly those little twigs turn into trees. And as well as being important in terms of conservation, don't underestimate the importance of the pure pleasure that woodland provides. The New Forest and the Forest of Dean are as valued for their leisure opportunities and their beauty as the magnificent formal gardens and landscaped parks for which Britain is famous.

Small gardens give tremendous pleasure at local level, and smaller woods, too, are vital in enriching our local environment. Enjoy this story of one such small woodland.

Preface

The idea of this book began to form two or three years after we bought the wood. This was partly because I had begun keeping brief notes after each visit, and it is these that I've been able to draw upon. The passage of ten years has yielded many events, surprises, and happenings from the evolving cycle of woodland life to make the story worth telling—at least, I hope so. I hope, also, to entertain, inform, and amuse as we all become more concerned about the well-being of our countryside. Few people will have the privilege we've enjoyed of owning a wood, but my aim is to share the fun as well as the vicissitudes of this experience.

In writing this book, I have had many kinds of readers in mind. As well as some knowing smiles from forestry colleagues, I hope farmers wondering how to manage a bit of woodland or whether to plant some set-aside will find here something to smile about. Indeed, I hope this will be so for anyone curious about natural history, rural goings-on, or the life of trees and woods in the English landscape. Perhaps, too, John White's and Stephen's illustrations, all of which are sketched from scenes in the wood, will help brighten a dark winter's evening, maybe when there's only television for company, and remind of the countryside with which we are so blessed.

Turning to a point of detail, imperial units are generally used in the book. The exceptions are the few references to tree diameters and volumes which are firmly metric today. But there is even an exception to this. As related in the book, when selling fine hardwoods like oak and ash one still uses the eighteenth-century system of 'hoppus measure'!

While this book could not have been written without the brief notes of each visit, the initial idea was prompted by an interview broadcast on BBC's *Farming Today* in early September 1987. At the wearily early hour of 6.15 in the morning, Margaret and I listened to an author on the programme enthusing about a book which he wanted to write 'for his wife to enjoy'. I think we remarked that there are quite a few books about life on a farm but none we knew of about life 'with' a wood. Since that early awakening eight years ago, and hoping that my wife would enjoy what I was writing, Margaret has read each chapter as it's been drafted. It has been a drawn out process and very intermittent as spare time for writing seemed ever more at a premium, but her remarks have not been at all discouraging—indeed, quite the opposite.

J.E.

January 1995

Acknowledgements

I am particularly grateful to the neighbours and friends and, indeed, to all who have been part of the wood's story, who have kindly read what I've written about them and not been too dismayed. Without them the story would be greatly lacking. In a way they almost are the story, although the wood itself is the hub and focus. All facts and events and people are true as I have recorded them. I apologise if at any point my interpretation has been mistaken, but I hope nothing I have written will give cause for offence.

John White, author and the Forestry Commission's dendrologist at Westonbirt arboretum, drew most of the splendid illustrations. Stephen, our middle son, beautifully sketched Melvyn's caravan in Chapter 5, the harvesting machines in Chapter 11, the sketches heading Chapters 1, 8, and 11, and the cover illustration although the idea for it was Benjamin's, our youngest son.

I am most grateful to Alan Titchmarsh for his Foreword as well as for opportunities to visit his own successful tree planting and adjoining woodland, not so different from the one in this story. Thanks go to Neil Middleton, Earthscan's book editor, for initial encouragement of the idea, and to Jenny Claridge for advice on style and what might interest readers. John and Elona Webber kindly read the typescript and my colleague, Roy Lorrain-Smith, carefully checked the proofs. Thanks are also due to the ever-diligent attentions of Oxford University Press.

Margaret, my wife, has known about this project from the beginning and her support and interest have been invaluable: I think she has even enjoyed it, too. We have prayed together about the wood and gratefully acknowledge the One before whom we do all things (Col. 3.23).

Contents

1

The purchase

This is the story of a small Hampshire wood. It is not the whole story, even if that could be told, but of the last ten years since we became its owners. It's a story, too, which touches on that of the owners themselves. Mostly, though, it is a more leisurely reply to the oft-asked questions: What do you do with the wood? Do you need to look after it? How did you come to buy it? Indeed, why ever *did* you buy one? We haven't got all the answers, and certainly not for every new experience our purchase has brought, but we've enjoyed finding that out since the wood became our own.

I don't think that buying a wood was my way of becoming a 1980s yuppy. Certainly, what we bought was on the market because of a form of privatisation, but it can't be compared with the instant profit from the great utility sales, which for us began and ended with the minimum allowable flutter on British Gas. Nevertheless, in April 1985 we owned a wood. It was not quite big enough at 22 acres to call a forest, though it contained many promising beech trees and fast-maturing pines, and in places was so thickly stocked to make it quite dark and dense. The acquisition, like all purchases really, had in the end been rather sudden although we had toyed with the idea for almost a decade.

My wife, Margaret, and I bought the wood in partnership with John, her brother. Almost ten years before, John had visited us for a few weeks in Papua New Guinea where I was lecturing in forestry at the University of Technology in Lae. One weekend during his stay we drove through the forest-drenched hills of that tropical island to Wau, pronounced 'wow', in the cooler highlands to escape for a few hours the unrelenting, equatorial heat and clinging, sapping humidity. On the way we talked about the idea of buying our own wood, I think because naming it 'Wau wood' appealed. Exactly how the conversation first started we've all long forgotten, and we never did call the wood after that New Guinea town. John had become interested in woodland after hearing Kenneth Rankin, the father of modern private forestry, talk with beguiling enthusiasm about such an investment as a mixture of fun, finance, and freedom from tax. From then on John was hooked on the idea or rather, as his aspiring partner, I hoped he was. Our circumstances, John a solicitor and myself a forester, seemed perfect if anyone was going to make forestry pay.

Most foresters don't own forest. Unlike farmers with their average 183 acres, or dentists their practices, or even solicitors their partnerships, few foresters have a financial interest in the woodland they manage. Unalloyed professionalism, untainted advice, unsullied integrity must surely be born of this, but so too the hope, even the longing, that one day they will have their own patch of forest. Of course, the compensations of a forestry career are many, but becoming an owner is rarely one of them.

Two circumstances led me into forestry. As a boy, evenings and weekends were spent playing in the National Trust woods at Petts Wood, now in the London borough of Bromley. They are a remarkable 180 acres of woods, streams, and fields in London's green belt just 12 miles south-east of Trafalgar Square, and are named after the Pett family who were shipbuilders in Tudor times, and who leased the woods for their oak. In the woods today stands a sundial commemorating William Willet, whose habit of riding through them on horseback at 7 a.m.—his most joyous hour—prompted the idea of daylight saving, first introduced on 21 May 1916 during the First World War, thus sowing the seed of the equinoctial ritual of clocks moving forward and back. Willet's birthplace of Farnham in Surrey has yet to honour his memory. Also standing in Petts Wood are remnants of great oak standards from the heyday of coppicing, such as the 'T' oak with two massive limbs sitting horizontally on a short, equally massive trunk, and lying prostrate is 'the crocodile', a fallen hollow log that small boys could crawl through and emerge nicely grubby. Only 40 years later did I learn that Margaret and John, who also grew up in Petts Wood, had similarly rejoiced over this log in childhood with its gaping mouth and inviting dark interior, and had also called it 'the crocodile'. Mucking about in the woods one learned that trees, like mountains, are for climbing. Finding that out, and even squirming through hollow oak logs, while hardly qualifications for a forestry career, certainly lent affection.

Interest was really aroused by a three-day forester's badge course for senior scouts run at Broadstone Warren on the edge of heath and woods that constitute Ashdown forest in Sussex. During the mixture of daytime practical work and evening talks by the light of a Tilley lamp, the realisation grew that a whole forestry industry was busy with growing trees for paper or planks. The appeal of the outdoors was a bonus, but from the moment of gaining the forester's badge my aim was to study forestry at university. In the biology sixth, my final year class at the City of London school, the news was greeted with quizzical looks from the rows of hopeful medics and vets destined for Oxbridge, Bristol, or London. Hardly any of them had even heard of Bangor in North Wales, where two years later I went to study forestry.

We had gone to Papua New Guinea seven years after graduating, and returned three and a half years later impressed by the diverse rain forests and enriched by the even more diverse peoples who can claim more than 700 of the world's languages.

On resuming work with the Forestry Commission at its southern research station, the thought of buying a wood was just one of many jostling for a place in the family's plans. Our son, Jonathan, was no longer the babe in arms we had taken to the Far East, and number two was on the way. Spare time was occupied writing my first book, about plantation forestry in the tropics. The effort rather confirmed my mother's observation that, while every husband might not have a sympathetic pregnancy, most tend to be unusually creative in the months leading up to their wife's confinement. I cannot vouch for the converse, unless Margaret has been keeping me in the dark while I write this! In 1979 Margaret's creativity was the more productive and our second son, Stephen, arrived well before my tropical book was ready. Indeed it was three more years in gestation, and even our third child, Benjy, had arrived by that time. But the thought of buying a wood was not entirely dormant, the appetite being whetted daily by research into how best to cultivate and promote the native hardwoods of oak, ash, and beech which so bless Britain's countryside.

Nearly three years before we finally acquired the wood in this story, we took the first serious look at one. It was called Birchen copse and was in the heart of mid-Wales. Three of us (Roger, another solicitor, John, and I) drove five hours to the loveliest of scenery to find a long thin strip of Douglas firs and larches which made up an 18 acre wood. We spent an hour looking over them and found the trees to be growing well, though many of the larches had swept butts, that is, the bottom six or seven feet of the trunks were curved like giant hockey sticks. The firs were a dark, dense mass and obviously held plenty of timber. They badly needed thinning, offering the prospect of early income. We walked across several fields to reach Birchen copse, a route which would have to be travelled in reverse by every log produced to reach a market.

We enjoyed the day, the wood was a size we could cope with, and I prepared a short synopsis of its prospects. We put in a rather half-hearted bid, partly perhaps because of the distance from Hampshire. It cannot have been the best offer and our excursion and brief dabble came to nothing. However, we learned a lot. We were much clearer now that the enjoyment of owning a wood is more than growing a commercial crop. This Welsh possibility was a long half day's journey away, meaning that picnics and camping, quite apart from informal management, could never be the impulse decisions so much the essence of leisure. In any case, if at the time we had gone ahead, two solicitors to one forester was perhaps unbalancing the scales of justice!

Interest in buying was fanned from time to time when copies of Clegg's particulars of woodlands for sale dropped through the letterbox. John Clegg and Company are agents specialising in the sale of woodlands, as are other estate agents such as Bidwells, Savills, and Strutt and Parker, and they may have 70 or 80 properties on their books at any one time. Most of what was on offer in the early 1980s seemed to be in Devon, west Wales, or Yorkshire and similarly distant places for us, and most were too big and too expensive.

The particulars which finally excited, in the summer of 1984, arose through a conversation at coffee with a colleague at work. He was talking about the woodland he looked after for his father and mentioned that he had been sent details of six small properties being sold by the Forestry Commission in north-east Hampshire. John Newcomb, a forestry consultant and chartered surveyor, was handling the sales. The woodlands were on the market because the Forestry Commission had begun disposing of some of its plantations to raise money and to rationalise its large and scattered estate. Over the years, but especially just after the Second World War, the Commission acquired numerous small woods both by direct purchase and as receipt of death duties paid to the Treasury. It was part of the Forestry Commission's task of rebuilding Britain's pitifully meagre forest cover, which had fallen below 5 per cent of our land surface at the turn of the century. Many of the smaller woods proved costly to manage, especially if far from the forest office, and it was sensible to market these first when the Commission began its disposals. It was sensible, too, since such woods proved popular with small investors, neighbouring owners, and local councils interested in improving amenities. We were one such small investor.

Three of the six properties for sale were outliers of Micheldever forest in mid-Hampshire. Their size, location, and asking price were close to what we wanted and, more importantly, John was still interested in buying a share. I visited all three woods one blustery autumn day. One was too small and was as inaccessible as Birchen copse, and of the other two, one was twice the cost per hectare though the crop didn't appear to be twice as good. The main recollection from that first, brief visit to the cheaper of the two woods was an impression of dense undergrowth, of areas

thick with pines, and of parts of the wood an impenetrable tangle, and that a more careful inspection was warranted.

John and I next looked round the wood together, in the ensuing weeks, followed by a trickle of relatives bemused by our intentions. One such early visit was on New Year's Day 1985. Snow had fallen the previous week. In the wood, trees had shaken free of their white coats, rabbit and deer tracks criss-crossed, and the air was crisp and clear. It was cold, but not biting, calm but not dank, a chilliness that a brisk walk would soon rectify. We set out on our tour of inspection and invigoration, and we met Alistair. Gamekeepers are a genre apart and Alistair was no exception. He was quite tall, but on this occasion what mattered was that he had just had a punctured tyre, had managed to fit the spare despite the numbing cold, and had then skidded in the icy conditions. Understandably, he was not in the best of humour. Of course, we didn't know that then and, importantly for us, Alistair didn't know what we were doing in the wood.

My mother and Marilyn, my sister, were with me. Marilyn had been delighted over the supplies of firewood the wood might afford and had enquired innocently if she could gather some there and then, as there was certainly plenty lying about. I had hurriedly demurred, pointing out that we were not the owners yet. Indeed, at the time, our first offer had been rejected and a second,

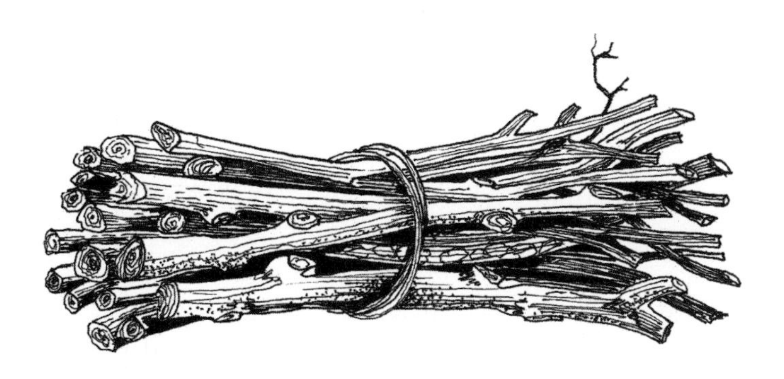

improved one was still being considered. Alistair's wrath might have been complete if not only were we trespassing in his pheasant shoot, and in season at that, but also caught clutching bundles of wood. Such a Christmas card scene, 'family gathering faggots', would have been picturesque, even idyllic—all that was missing was the tinsel and an overly rouged robin—but the circumstances were not.

We were walking back along the middle ride when we spotted each other, or rather Alistair saw his quarry. He was in the hunt, having seen our car at the entrance and doubtless our footprints in the snow. 'You know it says there is "No right of public access" in this wood?' That was all Alistair said. Not exactly a verbal assault, being a man of few words: the question said it all, and it was true, and he was right. At the entrance there was such a notice beneath the Forestry Commission's name surmounted by its crown, and the name of the wood: it reads 'Reserved rights; not open to the public'. At least there hadn't been a 'Trespassers will be prosecuted', neither the full nor Piglet's version of 'Trespassers W' which Winnie-the-Pooh would see, and there was not the bald offensiveness, still common in the countryside, of 'Private' or 'Keep out'.

Marilyn and Mother said nothing, withdrawing a little; I proffered the agent's particulars. Alistair's concern subsided, the weather came into the conversation, something was mentioned about the wood having been on the market for some time, and the day's catalogue of mishaps emerged bit by bit to explain the precipitate challenge. He volunteered an opinion on the woodland's price and threw in a few comments about the bark stripping at the base of some of the beech trees, suggesting that rabbits were the culprits rather than grey squirrels as I had thought. On reflection, this initially unpromising start was not such a bad thing. Here was someone checking the well-being of a wood not on behalf of the owner, the Forestry Commission, but the sporting tenant with his interest in pheasants. We later learned that Alistair visited the wood on his rounds on most days. And he had met others in the wood who were there with rather less good cause.

In February 1985 our offer, just below the guide price in the particulars, was accepted and contracts exchanged in a way no different from buying a house. This was John's territory and he looked after all the details; Margaret and I just signed the forms, though what our neighbours, Bob and Corrie, thought as they witnessed our signatures may or may not be revealed by their mutterings about capitalists and landed gentry. We were not buying the freehold, but a lease entitling us to grow trees and carry out all types of forestry work for the next 967 years, or from since the time of King Harold, thinking backwards to give a feel of the years at our disposal. The idea of a lease didn't worry us. Indeed, quite a few of the Forestry Commission's woodlands are owned as a long lease that conveyed everything needed for pursuing sound forestry while keeping for the freehold owner usually just the rights to sporting.

The wood was just under 22 acres, which sounds more impressive than 8.7 hectares, and we bought it for the price of a good quality family car. This worked out at something like a quarter of the price per acre neighbouring farmland would have fetched. John and I split the cost 50:50, but we both think of ourselves as owners of 22 acres. Our share amounted to a kind of financial recycling, being more or less equal to the royalties received in the three years since the book about tropical plantations had been published.

On 4 April 1985 the wood was ours. On 8 April we applied to the Forestry Commission for a felling licence. This was not to lay waste what we had just bought, a step well-nigh impossible today without exceptionally good reasons, but to allow thinning and cleaning out of the scrub birch, sallow, hazel, and poorer beech and pine. On 30 April the application was rejected. We had been too hasty. The local forestry office had not been notified of the change of ownership and from their point of view the felling licence application appeared like asking to cut their trees in their wood for our convenience and our profit. It was not as melodramatic as that, but the staff handling the matter must have wondered what was going on. The misunderstanding was soon sorted out. At much the same time, as if to confirm we really were the new owners, the shallow triangular sign board at the entrance bearing the Forestry Commission's name was retrieved by the Commission along with the padlock.

Buying a wood is really like buying anything else. You like the look of it, the price is right and there appears nothing to put you off. With a wood there are one or two important things to watch out for. First, though, Margaret and I prayed over the idea, not only about the rightness of such a purchase, but that if we bought it we might find opportunity to share it too, in the widest sense. What was the wood like and what had we looked for when buying it, in addition to satisfying ourselves about the trees as a timber crop?

First, the wood was not too far away. In fact it was almost midway between John's home and ours, giving us both a journey of about 12–15 miles. Secondly, on the west side the wood was bounded by a lane and on the east by the London to Southampton railway line. This was important—the lane not the railway—because it is all very well owning a woodland but it will be of poor productive value if it is remote and inaccessible. Much of the cost of getting timber from tree stump to mill is dragging the logs to where a lorry can be loaded. If this is difficult, as it might have been with Birchen copse in Wales, the grower's modest returns are further diminished. The access from the lane was good since some years before we bought the wood the Forestry Commission had constructed a wide bell-mouth entrance which opened to a short track and turning bay with solid formation and metalling adequate for laden timber lorries.

Access within the wood was good, too. The short, well built entrance track gave way after 30 yards to an informal track which ran through the wood to the railway where it forked and continued to both northern and southern boundaries. Under most conditions these tracks are motorable and do not rut badly: the soil is well drained and puddles are rare. Half way along the main track, it is bisected by a grassy ride which runs across the wood on a north–south axis. Thus the whole block is divided into four almost equal squares of 5–6 acres each. Originally these were probably four fields since remnants of hedgerows can be found beside both the main track and the cross rides. The occasional guelder rose, wild privet, field maple, wayfaring tree, and other common hedging species are still present, often with a few violets and primroses too, marking where the old field margin once was.

This excellent standard of access strongly influenced our decision to buy. Added to this, the wood was virtually square in shape, which is a lot easier to manage than a long thin strip, and the ground surface was even. The whole wood sloped gently with an easterly aspect from a high point of about 420 feet to 385 feet beside the railway, and it was not too exposed. Also, we cannot deny that the numerous clumps of primrose leaves, some well advanced and some still like tiny fleshy rosettes, seen emerging in one of the cross-rides in that first spring, beckoned of the splash of yellow to come, and warmed our interest.

All these considerations apart, we still had to take the plunge and buy something whose assets were not readily realisable. There is no doubt also that for the first few days of ownership we, or at least I, wondered if it was going to be all right in our care, just like when mum and newborn baby are home from hospital and you tiptoe to the cot to see if the mite's still breathing. The fact that the wood had survived perfectly well for years without us was no solace to the irrational worry about its well-being. I did pop over to see it rather frequently in those early days; but worries faded, affection grew, and the rewards long outlasted and greatly exceeded in pleasure the small capital gain on British Gas.

2

What we had bought

When we bought the wood it had three different parts. First, beside the railway was a 4-acre strip of assorted deciduous trees, mainly sycamores and scruffy oaks, and some conifers including spruces, pines, and quite a number of enormous larches. We didn't own these trees, which were retained by the lessor from even before the Forestry Commission's ownership. The lease permitted this arrangement to last until the year 2003, by which time they had to be felled and the land come into our possession. Rusty wire and an occasional rotting fence post marked the boundary between this strip of old trees and the rest of the wood.

The second, and larger part of the wood, amounting to about 14 acres, consisted of 28-year-old trees which the agent's particulars said had been planted in 1957. These trees had the main commercial potential—the crop in farming parlance—and consisted of alternating three row-wide bands of beech and Corsican pine. This paints pyjama stripes of plantation on steep hillsides, but our wood is gently sloping and does not intrude into the landscape. Only the microlight enthusiasts, who forever whirr overhead from the nearby Popham airstrip, could see the wood striped in this way, especially in October when the reds, russets, and bronzes of autumnal beech contrast strikingly with the sombre green of the pines.

The remainder of the wood was three irregular patches of trees each with a few Douglas firs. This is a splendid timber tree from north-west America where it grows to 250 feet or more, and is named after David Douglas, the botanist and explorer who first brought seed into Britain in early Victorian times. When crushed, its inch-long, dark green to bluish-green needles exude a fruity delicious aroma of tangerines. Besides the Douglas firs there were what seemed a motley collection of beech, sycamores, and occasional oaks which had been lumped together in the sale particulars as low-grade broadleaves, the coniferous Douglas firs notwithstanding.

I find, like so many people, that only after a purchase does one really examine intently what has been bought. It is the wrong way round, and one tries to guard against this potentially disastrous habit by employing surveyors to inspect houses before exchanging contracts or the Automobile Association to check over a second-hand car. Somehow, though, the excitement of the purchase prejudices objectivity—the 'what on earth does he see in her?' syndrome! The woodland was like this; initial impressions suggested an attractive proposition, but closer inspection revealed how much a mixture of tree species it was, the variety of ages and quality, and some unexpected features.

First, very obvious, and not unexpected was that every third row of pine trees had been cut in a thinning some years before. This will have earned a little revenue and provided more space to help the remaining pine and adjacent beech to grow. Walking down the main track one can look along these regular 'alleyways' and, in winter when leaves have fallen, see clearly to each boundary. They make superb sightlines for spotting deer moving through the wood. The trees were planted in straight rows running south-west to north-east. Indeed, on 27 November, and presumably the analogous date in mid-January, the setting sun shines along their entire length. While not matching for age or symmetry what so impresses the solstice watchers on midsummer's day at Stonehenge, it is a remarkable testimony to how straight a line was planted, and is now revealed by thinning, and illuminated with theodolite precision by the rays of a November sunset. The second thing we noticed was that the beech trees in the mixture, and the euphemistically named low-grade broadleaves, were neglected and untouched: wild *Clematis* was everywhere. This vigorous climber merits at least a paragraph; it can be a countryside menace.

Wild *Clematis* can be found swamping a hedge or curtaining the edge of a woodland wherever there is chalk downland. It is graceful enough around Christmas when bedecked with baubles of greyish-brown seed heads and aptly named 'Old Man's Beard', but is otherwise known as 'Traveller's Joy' and sometimes as 'Virgin's Bower'. Why it is called 'Traveller's Joy' is difficult to understand since it habitually renders young woodland or thicket impenetrable, cobwebbing together every stem and limb. From a distance it can appear draped, resembling old snow, grimy, pale grey, on the point of thawing. And, for some people, the sap of *Clematis*, which like buttercups contains ranunculin, irritates the skin, raising blisters. The creeper can grow and thicken unhalted for years, becoming like a tropical liana, and its rapid upward assault on a tree can quickly elevate it 30 or 40 feet, and sometimes even more, into the branches of the crown. Unrestrained, *Clematis* will sprawl everywhere and overwhelm, even physically by sheer weight, with only ash of all trees seeming able to grow above it unhelped. In New Zealand, where it has been introduced, it is even worse, appearing rampant everywhere and not confined to chalky soils. In Britain its redeeming features are that it supports two attractive moths, the Small Emerald and the Clouded Chalk Carpet, and, of course, the very tangle it weaves is marvellous cover for woodland birds such as warblers and finches. There is an explanation, too, for its common name. Keith Kirby, English Nature's foremost woodland ecologist, told me that the herbalist Gerald, writing in 1597, interpreted its medieval name of 'viorna' as 'viorna quasi vias ornas'—'of decking and adorning waies and hedges'; and since we travel along ways and hedges, he coined the name 'Traveller's Joy'.

The large amount of *Clematis* in the wood was one sign that cleaning of the plantation had been incomplete some 15 years before. Another indication of this was an abundance of sallows, birches, thorns, and briars which spring up everywhere land is abandoned, and which in places in the wood were well outgrowing the beech. Cleaning is a time-consuming job in forestry, tiresome and costly with little to show, but often essential if the planted trees are to succeed in becoming the main crop. It's no wonder that with finances tight, economies had been made in cleaning the wood. This decision wasn't, silviculturally, such a serious shortcoming. After all, 15 years later the plantation was sufficiently well stocked with pine and beech trees to make it attractive enough for at least one purchaser.

Cleaning was the first forestry skill I learned in the school of experience. My first paid job, as an aspiring student in the summer of 1965, was weeding and cleaning in Forestry Commission woods administered from its old Badgers Mount office near Shoreham in Kent. On the first day I was handed a 'slasher', an appropriately named implement which is a sort of hybrid between a sickle and scythe. The curved blade is quite short and attached to a three-foot-long wooden handle. Armed with these tools the gang, with its novice, set off walking, first beside a young mixed plantation of oak and spruce, on down a sloping field, across a valley, and then up to the 15-year-old beech wood in need of our attention. Today the mixed plantation is grown up and shelters the car park and picnic area of Andrew's Wood, and the lovely dry valley we had to cross is now the M25. The motorway cuts through the bottom of the beech plantation we cleaned, though most of it still remains on the south-east side. In those less hurried days I was set to work next to the ganger using the slasher to hack down all unwanted growth, whether briars, thorns, foul smelling dogwood, or climbers such as wild honeysuckle and *Clematis*. The knack is to wield the slasher rather like a golf club and slice through the woody stem just as the blade begins its upswing. Every exertion of youthful enthusiasm on that May morning, doubtless fired by the new boy's naïve wish to impress, only yielded half the progress and over twice the number of cuts and scratches of wise experience plugging steadily away up ahead. We slashed and hacked in that wood every day for two weeks. Sores and blisters were one introduction to the world of work, and slashing at *Clematis* to unstrangle whippy shoots of beech, which slapped the face and stung the eyes in defiance, has left little affection for the creeper and a stoical appreciation of this irksome task.

Growing pine and beech together in narrow, three-row bands is a successful mixture and quite common on the downs and clays of southern England. The pine is steadily thinned out and finally removed, usually well before it is 50 years old. This earns some revenue while the beech left behind have space to grow on for twice or three times as long until large enough to yield the creamy-pink furniture timber so much admired. The pine in the wood is Corsican, originally from the island of that name, and is not native to Britain. It is a close relative of the heavily branched, black-barked Austrian pines beloved of Victorian railway builders and now, in their old age, a more enduring memorial than many of the country stations and sidings beside which they were planted. Much Corsican pine has been planted since the last war in the warmer and drier south and east, being generally superior in growth and straightness of stem to our native Scots pine. More important for our wood is that it grows better on chalky soils and is probably why it was chosen for planting.

Beech is native to Britain, but not to all parts. Despite the many fine stands in northern England and Scotland, all were planted by earlier generations. The species is thought only to occur naturally in southern England and as far west as Gwent. It is difficult to trace this: analysis of pollen in ancient peat beds and the composition of charcoal fragments from Bronze Age fire sites provide the main clues. But much confusion has been wrought by countless unrecorded introductions from across Europe. Sometimes seed of very good trees has come in this way, such as the large quantities of beech nuts soldiers stuffed in their pockets when returning victorious from Belgium and the battle of Waterloo. These unusual spoils of war from the Forêt de Soigne have grown up into some of today's finest and straightest beech trees.

The beech in our wood are from seed collected from Savernake forest in Wiltshire and exhibit the characteristic variability of this mediocre source. Some trees are twisted, almost snake-like and poorly formed, some are very vigorous, others more like runts. On the whole enough are straight stemmed to promise a fair crop. We certainly hope so. Our grandchildren, not ourselves, will mostly benefit and they will see whether this expectation is right. At least William Cobbett, that early nineteenth-century farmer, politician, and itinerant, would have approved of growing beech in our part of the country. Eulogising in his *Rural rides*, Cobbett marvelled: 'This lofty land of north Hampshire, the finest beechwood in all England, large sweeping downs and deep dells, with villages among lofty trees'. Though even impressive trees didn't take pride of place, as he later exclaimed: 'What in the vegetable creation is so delightful as the bed of a coppice bespangled with primroses and bluebells?' What indeed?

As John and I made our first visits to the wood in 1985, we began to notice between some of the rows of pine that the beech were dead or missing altogether. The dead trees and many others nearby, though still alive, had been partially or completely 'ring-barked' just above ground level some years before. Two animals gnaw or strip bark in this way: rabbits and grey squirrels. When their exertions encircle a stem—ring-barking—the vital link between crown and root is severed and the tree soon dies. Our culprits, as I learned from Alistair and saw later for myself, were rabbits. The wood was heaving with them. And, with one long boundary next to a railway, we could hardly expect anything else, knowing the legendary attraction of British Rail's embankments. When the present trees were planted in 1957 myxomatosis effectively controlled wild rabbits. Today numbers are nearly back to their former level. Indeed, the Forestry Commission now reports rabbits present in 200 of its 204 forests.

The rabbit is often quite a serious pest and, not surprisingly, it was never originally part of our wildlife. Mediterranean in origin, the rabbit was domesticated by the Romans and probably introduced to Britain for food by the Normans who farmed them in fenced enclosures, called 'warennes', hence the name of their underground home today. They soon escaped to become a nuisance to farmer and woodman, though until the eighteenth century were doubtless kept in check by foxes and wolves, by raptors such as the kite and buzzard, and by the less well off people in those less squeamish but more hungry days. Fortunately, they never caused in Britain anything like the devastation which followed their hapless introduction into Australia. Just 100 years after their arrival in the antipodes, Australia was awash with rabbits. Herculean efforts were made to contain them; the state of Western Australia alone tried to fence them out over a distance of more than 1000 miles, but to no avail. Uncontrolled by predators, the bobtail tide browsed foliage and stripped bark to ruin habitat after habitat, both for themselves and for the much loved native marsupials such as wallabies, kangaroos, and koalas. In England they have never been a menace on this scale.

A rabbit-infested wood is readily identified by the look of neighbouring fields, particularly if planted down to wheat, barley, or a legume such as peas. The height of the crop diminishes towards the woodland edge, a bit like Concorde's tapering nose, when rabbits are in residence. It explains why the law requires owners to control them. They browse most near woodland, but can damage a farmer's crop 20 or even 30 yards into the field. This same browsing habit is what harms young trees in their first few years. Either rabbit-proof fences or individual tree protection is essential today in most parts of Britain. Despite the enduring appeal of Beatrix Potter's Peter Rabbit, the forester usually has to side with Mr McGregor.

In our wood the ring-barking of beech trees almost certainly occurred in the winter of 1981. Between December and late February snow lay on the ground for long periods. With everything blanketed and hidden, beech and sycamore bark was the rabbits' last source of sustenance. Little of this stripping damage has occurred since, at least not by rabbits and not until 1992 when the other grey exotic, grey squirrels, began to do so in earnest. They have a chapter to themselves later on.

At least 100 beeches appeared to have been killed or severely maimed by stripping bark, including a few trees which were over 30 feet tall and as thick as a gatepost at ground level. Such damage is not, of course, confined to our wood and the yellowing and dieback of many beech trees in southern England has been due to rabbits and squirrels and the droughts of 1976, 1984, and the 1989–92 period. It is worth recording that February 1990–April 1992 was the driest $2\frac{1}{2}$ years since the 1850s.

As well as rabbit damage to the beech, we noticed that patches of them and of Douglas fir, and even some pine, were sparsely crowned and sickly. Leaves and needles were pale, yellowish, a little mottled and distinctly off colour. The malaise is labelled chlorosis. The severity of the great drought of 1975 and 1976 probably triggered this problem since it is now clear that many large beech on shallow soils failed to recover afterwards, lingering on for some years with few leaves, thinning crowns, and chlorotic condition before finally succumbing. Long-term studies show that previously healthy mature beech, well short of old age or tree senility, were arrested in their growth by the great drought, to stagnate and die in the years that followed. Twigs and small branches of beech easily reveal the extent of each year's growth: a bit like annual rings in wood, a narrow band of fine wrinkles in the bark demarcates each year's extension. The beech in our wood are not yet old or large, but occasional patches of sickly trees and the poor condition of some Douglas firs pointed not only to past droughts but a more deep-seated cause: an excess of chalk in the soil. The soil throughout the wood has chalk fragments right to the surface as well as chalk rubble at depth as the underlying geology. Pouring a little dilute acid such as household vinegar on the soil makes it fizz because of the chalk. This degree of chalkiness denies some plants key nutrients, especially iron and manganese, so causing the chlorosis. It is the same problem gardeners face when wanting to grow heathers or azaleas in the chalk and limestone regions of southern England. Unfortunately the gardening solution of applying sequestrene to add the missing nutrients is hardly practical for a whole wood! Instead, we have to put up with the best of natural indicators of chalk soils—*Clematis*.

Our soil is so very chalky for three reasons. Before the wood was first planted—we believe in the 1890s—the land may have come under the plough. With more or less pure chalk at about one foot depth, and as easily rootable rubble, quite a bit would have been brought to the surface during cultivation. Once in topsoil, chalk takes centuries to dissolve. It is also possible that the land was limed while it was farmed in Victorian times, and lime is finely ground chalk. The third reason is rabbits. Their warrens and numerous burrows all over the wood are as effective as any plough in turning over soil and bringing chalk to the surface. In places, they were further helped in this mixing by moles. Thus our soils are classified as calcareous or chalky and trees growing on them prone to chlorosis.

The bulk of the trees are healthy and growing moderately well. It is only 'moderately' based on the grades of tree growth rate the Forestry Commission calls 'yield classes'. Compared with the national average, the pine and beech are just a little better but far from being anything exceptional. The wood contains many thousands of trees, and in 1987 the pines were counted, with the help of my mother (Nain to our boys), Jonathan, and Stephen. There was a total of 2118 trees over 10 cm in diameter (this is the metric cut-off point of what used to be four inches). We measured every tenth tree and discovered that their average diameter was just over 20 cm, or about 8 in., at 1.3 m above ground level—an internationally agreed position for measuring trees equivalent to 4 ft. 3 in. (4 ft. 6 in. is used in America) and rejoicing in the name 'diameter at breast height' or DBH for short. The diameter of the largest pine was an astonishing 38 cm, that is, more than 16 in. Heights were not measured accurately, but the tallest were about 50 ft. or quite a bit taller than telegraph poles. These dimensions showed that the pine would be ready to cut for sawlogs in a few years. The beech were generally smaller and more numerous, never having been thinned, perhaps 5000 in all. The number of Douglas firs was about 120. We did not count the low-grade broadleaves. In all, the wood was stocked with about 9000 trees of all sorts and sizes.

Putting these data another way, the total weight of wood in all the trees amounted to rather more than 1000 tons. That was what Margaret, John, and I bought and it was being added to each year by about 80 tons since the trees were at their most vigorous: truly a growing asset.

Buying the wood was, of course, more than just buying a collection of trees. We also became proud owners of a large grey water tank, sited in the middle at the crossroads of the ride and main track. Fire has long since passed as the main threat but the tank has remained sound and full to the brim with murky water. The boys have stirred it with sticks, but some things are really better left well alone!

Along the southern boundary is a hedge. At least it ought to be a hedge but is really no more than a line of overgrown hazel with an occasional scraggy ash and sycamore poking through. The hedge has never been laid, indeed hazel isn't particularly amenable to this rural craft, and every year it just gets a little bigger.

The Forestry Commission's fence, put up 30 years before, has long since fallen down, with an odd stake and bit of rusting wire as all that remains. Since the neighbouring field was used for cereals, and provided we kept on top of the rabbits, there was no need for a formal barrier.

So we bought a lot of trees, a lot of burrows, a lot of work, and an awful lot of time. The timelessness of woodland life conveys something of Peter's ineffable statement about God: 'for with Him one day is a thousand years and a thousand years one day'. We have got nearly 1000 years on paper, but only perhaps 40 or so, God willing, for John and me to get down to work. With the life of a tree crop often longer than one's own, what is done is there for posterity. In our wood there was much to be done to turn a promising start into a fine crop. Learning forestry from lectures and textbooks is turning out to be a poor substitute for the real thing. I should have realised that from the very first day of cleaning a wood when *Clematis* and wise experience both got the better of a teenager. Perhaps if I had, what follows would have been spoiled. But first there were more unexpected discoveries.

3

Our plan to have a nice black frost for walking to Whitchurch and there throw ourselves into a postchaise ...

Jane Austen anticipating a walk from Steventon,
30 November 1800, to savour again
'one of the joys of heaven' — the beauty of landscape.

Mainly about railways, writers, and wildwood

The wood is squashed between a railway line and a lane. Its nearly square shape, and neat internal subdivision into four quarters, suggest it is of recent origin rather than a wood which has existed since medieval times or earlier. The coming of the railway may have led to the original planting. In the golden age of railway building the line at the bottom of the wood effectively cut off farmland at the western extremity of the sprawling Steventon estate. With access impaired (even now the link, Waltham lane, can only boast a narrow bridge with 11 feet 1 inch headroom), it seems that four fields were eventually put down to trees to form the wood. Obviously trees do not require the same daily attention as farming, making it a sensible change of land use.

The railway is the London to Southampton line, begun in the mid-1830s. It was fully opened on 11 May 1840 when the final Basingstoke to Winchester stage was completed. This is the section which runs past the wood and it was the last to be constructed, being just north of Litchfield tunnel, where the line reaches its summit of 390 feet and a distance of $55\frac{1}{2}$ miles from London. Parliament's originally sanctioned route for the line southwards from Basingstoke, proposed in the 1830s, was shorter than the more westerly sweep it now takes past Steventon, but included a mile long tunnel under Trinleys wood and Popham Beacon and an even higher summit. Even so, what was finally constructed required almost continuous embankments and cuttings, which bless the traveller with beautiful views of north Hampshire's rolling downs, but was a formidable task of blasting, digging, and heaving. More than 3.2 million cubic yards of soil and rock were excavated by the navvy gangs to furnish the even road and gentle gradients so sought by early Victorian engineers with their primitive underpowered locomotives. One in 250 was the most they could manage and, remarkably, such a gradient is maintained for 17 miles in this middle stretch. Joseph Locke was the engineer for the line and Thomas Brassey, who became the greatest and perhaps the fairest of all navvy contractors, cut his teeth on this final Basingstoke to Micheldever section with its massive excavation. Railway buffs marvel at 'Locke's stupendous cuttings and embankments between Basingstoke and Winchester'. Brassey's original 1100 men, whose loyalty subsequently took them on to the Paris–Rouen line[1], where they were joined by 5000 other British navvies he recruited, and then ten years later, to heroism in the relief of Sebastopol in the Crimea, may have camped where our wood is now. Where the line passes the wood it is level with it, neither in a cutting nor on an embankment, and hence a good place for the navvy gang to sojourn—though who supplied each man's daily ration of 2 lb. of meat, 2 lb. of bread, and 5 quarts of ale is anyone's guess. The longer route which Brassey's men built from Basingstoke to Winchester is the most 'remote' in the southern region, if remoteness is measured by inhabitants, with just one station, Andover Road or Micheldever as it soon came to be called, in the whole $18\frac{3}{4}$ mile run.

Next to the railway was the reserved timber and, to begin with, John and I were at a loss to account for this 50 to 60-yards-wide strip. When we bought the wood there were many large trees on this strip, as has been mentioned. We did not own them, but when

[1]Astonishingly, three-quarters of all the railways built in France up to 1850 were the work of British navvies and British money accounted for two-thirds of the capital invested.

they were felled in February 1987 by the lessor, Prudential Insurance, their stumps revealed about 75 rings, suggesting they had been planted in about 1910. This was at least 15 years after the bulk of the wood was first planted in the 1890s. Alistair, once again, was forthcoming with an explanation. It had been initially left unplanted because it was still used as a coal yard! Although local passenger traffic did not warrant a station, with the nearest some four miles south, goods trains could stop to offload coal for the village of Steventon about a mile away, and perhaps the hamlet of Litchfield too because the railway was level with the ground at this point. In the early part of the present century a siding was laid and local people can still remember it in amongst the big larches. For the years when Britain's empire was at its apogee and Victoria ruled unchallenged, even a village of a few hundred like Steventon could, when close to a railway, share in the Industrial Revolution's black gold and get its coal.

Steventon's coal was unloaded and piled beside a cart track which runs diagonally away from the railway to the southern edge of the wood. Beside this track one can still find lumps of coal and a lot turned up when digging the soil for planting young trees in the spring of 1987. At the edge of the wood the old cart track turns west and can still be followed beside the boundary hedge of tattered, overgrown hazel, all the way to the lane at the top. Thus was the route from coal yard to lane. From there, turning right along what is now the top of the wood, and right again at the crossroads brought the loaded cart to the thoroughfare of Waltham lane. This lane, hemmed in by hedges of thorn, hazel, and wild rose with an occasional emergent yew tree, descends steadily, squeezing through the low, narrow railway bridge, before it eventually climbs up a little and levels out coming to the next crossroads. There is barely a bend in this stretch of nearly a mile; indeed Waltham lane is as direct a route as could be surveyed between North Waltham and Overton, forgiving the odd wiggle approaching each village. At the crossroads the carts, weighed down and dirty with coal, would make a final left turn past Bassett's farm into Steventon. Neither weighed down nor dirty, but undoubtedly from time to time taking from Steventon this same route, would be the village's most famous daughter, the young novelist Jane Austen.

She was born in Steventon rectory and spent the first 25 years of her life living in the village before moving to Chawton near Alton. Both *Pride and prejudice* and *Northanger Abbey* were written while at Steventon. The attentive reader will already be remonstrating that Jane's life hardly overlapped the coming of the railways as she died eight years before the first public railway, Stockton to Darlington, opened in 1825. Indeed, the changes the construction of the London to Southampton railway must have wrought on her beloved landscape would, had she lived to today's expected three score and ten, undoubtedly have found expression in her novels, perhaps attributing to it the same mixture of distaste and gratitude as when a carriage failed to arrive to take her to an unwanted ball: 'disconvenience and disinclination go together.' Jane loved the countryside and rejoiced in the landscape, considering its beauty to be one of the joys of heaven. She liked a good walk, preferring cold, crisp weather to warmth, and wrote with relish of the prospect of walking to Whitchurch—almost certainly by way of Waltham lane and via Overton—provided there was '... a nice black frost!' Her dislike of warmth and humidity, just like my mother's, yielded in September 1796, in a letter to her sister Cassandra, to whom she was devoted, just the confirmation needed by all who wistfully speak of the real summers of bygone years: 'What dreadful hot weather we have, it keeps one in a continual state of inelegance!' So, with her proximity, zest for walking, a love of landscape, and incidentally, a family dedicated to country sports, I rest my case that the young lady, in the closing years of the eighteenth century, on occasion walked or took a carriage or postchaise along the lane at the top of the wood.

For Jane Austen there would have been no railway bridge to go under and no woodland where ours now is. All she saw were probably four fields of rough pasture, hedges and an occasional hedgerow tree. Old maps mark two tree stumps where great veterans must once have stood. The view would be of more interest as Jane looked east over the Steventon estate, to the gentle rise and fall of the north Hampshire downs. Looking west from the lane and away from the top of the wood, Jane would see across the field the spreading crowns of great standard trees of oak and ash in Bramdown copse. This wood is ancient and quite unlike ours with its recent Victorian pedigree. It long pre-dates Jane Austen.

'Ancient' woodlands are more than 'very old'. The term describes woodland which is known to have existed continuously as forest since 1600. And, so the argument goes, if it was woodland then it almost certainly always has been. Thus such woodlands are a link with the great wildwood or primeval forest that covered most of Europe in pre-Bronze Age times. No such woodland in Britain now remains wholly untouched and entirely natural. The nearest, extensive such wildwood is probably the Forest of Bialowieza in Poland. All the 730 000 acres of ancient woodland identified in Britain have been coppiced or managed in other traditional ways. They probably owe their survival to being useful and, because they were not neglected, they were not cleared for farmland. Most woodlands shown on the famous 1810 'first edition' Ordnance Survey maps are ancient, there having been little planting before then. Bramdown copse is on these maps; our wood is not—the map is just blank. Indeed, the wood only makes its cartographic debut in the 1911 edition of the six-inch series. Several new plantations are also depicted for the first time on this edition including the aptly named 'Railway plantation' next to where the line emerges at the south end of Litchfield tunnel.

Bramdown copse, or coppice in its unabbreviated form, with its standard trees is a relict of the commonest system of woodland husbandry up until the present century. Any tree species which throws shoots from a freshly cut stump, and most broadleaves do, can be handled so that these new shoots form the next crop. Normally such shoots, called coppice, are not grown on to timber size, but cut every 10 or 20 years for sticks, poles, or posts. Bean poles, pea sticks, and hazel thatching spars are still obtained in this way, as are chestnut palings from the coppices of Kent and Sussex. Often, not all the crop is kept as coppice and a scattering of trees are encouraged to grow to timber size. These are the 'standards' which tower over the coppice, or 'underwood' as it is called, like oaken sentinels patiently watching over the young regrowth. Oak is the usual tree for a standard, although ash is quite common. Beech, which casts dark shade, is unsuitable. All copses, or 'coppices', were once worked in this fashion, either with or without standard trees. In medieval times over half of all lowland woods were coppiced. As well as being anxious about his wives, Henry VIII was anxious about the nation's timber and passed a law requiring 12 standard trees to be grown in every acre of coppice. Timber for the navy was in short supply and capsizings like the Mary Rose didn't help. Such a fine ship of the line required 2500 tons of oak and the clearing of 50 acres of forest.

The interest in ancient woodland, beyond the purely historical, is that their link with the great forests of old survives in the richness and variety of the wildlife they possess. The famous forest of Alice Holt, straddling the Surrey/Hampshire boundary, and the Wealden edge hangers—the precipitous wooded scarps of east Hampshire—are good examples. Their poor soils, which probably protected them from being cleared for farmland, boast over 300 different kinds of plants and flowers not counting mosses, lichens, and ferns. New woods planted on farmland, or even ones developing naturally on waste ground, rarely have 50 or more. Indeed, woods planted as long as 200 years ago are still considered 'recent' and are impoverished compared with their venerable cousins of greater antiquity. Fortunately, Hampshire is rich in ancient woodlands. What appears important is not that individual trees are of great age, but that woodland life has continued for millennia, allowing woodland-loving plants to establish and maintain themselves through the long cycles of light and dark. This is why ancient woodland is of such conservation interest and why today their coniferisation (a term rather more unfortunate than its effects) has long ceased to be policy, despite the loss of much needed timber production. Fortunately, too, it now seems that even 30 or 40 years of conifers is not quite as harmful as once

thought. Rare woodland plants like oxlips have reappeared, even after decades, following felling of the conifer crop. What matters most is that a site is an ancient woodland one, not so much the type of tree cover at a particular time.

The richness of plant life in ancient woodland is one way of identifying its long pedigree. As mentioned, oxlips are exclusively woodland inhabitants, colonising new ground very, very slowly. If they are present, the place has long been woodland. Other good indicators are wild service trees and the herbs yellow archangel, wood anemone, and Solomon's seal. Dog's mercury was also thought to be a good guide, but this isn't always so, and both ancient and recent woodlands on chalky soils can be found, like ours, overrun by this interminable monoculture of green. Bluebells, too, are now discounted as a sure sign of venerable ancestry, and there is nothing to stop an owner deliberately introducing this attractive and peculiarly British wild hyacinth into a wood. We have planted three patches (there was none when we bought the wood in 1985) thanks to Ruth, my wife's aunt, who one day phoned with the urgency of a just-remembered birthday to say she had some bulbs from her garden and did we want them? Their rarity still in our wood makes them attractive to roe deer with their fussy appetites and possibly rabbits too, ravenous for greenery in early spring. Bluebells colonise slowly and it will be at least a couple of centuries before the wood can boast a sea of blue in early May like so many nearby. We have tried scattering bluebell seed, with little luck, even knowing that it takes four years for flowering to start. There have been no signs of green shoots where the seeds were sown: perhaps they were sterile, or gathered when still immature, or possibly the tender leaves simply proved too irresistible to browse. Ruth's kind help is the best approach, though one can't go digging up clumps of bluebells from anywhere. The 1981 Wildlife and Conservation Act forbids such pillaging of the countryside, not to mention the questionable purpose of embellishing a recent woodland with a few of the features of one more ancient, which people happen to find attractive. It will confuse ecologists in the future and confuse the natural occurrence of plants today.

What we are not confused about is the fun of owning a wood and the spreading of this enjoyment to our immediate families and beyond, even to gifts of bluebell bulbs from in-laws. For this we are grateful, and indeed that such things happen is an answer to prayer.

Only yesterday, as I write this chapter, Margaret and I were showing the wood to Bill and Norah from South Africa. In the course of the tour, a standard figure of eight for first-time visitors, we passed the enormous oaks with great shaggy tops that guard the adjacent woodland on our northern boundary. The largest would have been big enough to conceal King Charles escaping from the battle of Worcester, but it hides instead a murky pool in the crotch of the great limbs of this ancient pollard. These old timers, propped up by dense growth of sycamore, ash, and hazel coppice, probably guided the forester in deciding what to plant our area with back in the 1890s. Ignoring the post-war beech, pine, and Douglas fir of the Forestry Commission's fiefdom, there are occasional large oaks and sycamores, about 30 in number, scattered through the wood and plenty of uncleaned remnants of hazel. This suggests that when the Victorian fields were turned over to forest, coppice with standards was the aim, with hazel as underwood and a mixture of oak and sycamore as standards. And, if this is woodland that was cut in the war, the Forestry Commission's subsequent achievement of obtaining 90 per cent stocking of pine and beech from amongst the jungle of regrowth which confronted it is credit indeed in the days before herbicide.

Cutting overmature hazel is tiresome. Although the wood of hazel is not hard and the stems never too thick for even the smallest of chainsaws, they splay out in all directions denying access to the centre. As they are cut their outward lean pinches the saw and the tops, twisted and whippy, inevitably become tangled. Once cut, the stumps resprout vigorously, unless heavily shaded, and quickly restock the ground as is the way of coppice. Today, the maturing pine and beech have weakened the remnant hazel, but the large overmature clumps, gross in appearance, remain a problem. They are much too old to yield worthwhile sticks and poles, their firewood value is paltry, and the cost of clearing steep. An attractive option would be to return the hazel to a 7–10 year cutting cycle. Hampshire is short of such hazel in good heart in the midst of a surfeit of overgrown and ageing coppices from a generation of neglect. Interestingly, there are more traditional coppice craftsmen in the county skilled in working hazel to thatching spars and hurdles than good crops to keep them in business. It would be good if our wood could help ease this shortage.

The trees planted to create the wood in late Victorian times were cut as part of the war effort. Britain's private forests bore the brunt of timber supply during the blockade, the fledgling Forestry Commission, just 20 years old in 1939, having few plantations in production. It was the second time in just a quarter of a century that the nation's meagre reserves were called upon because of a world war, and it is not surprising that building a strategic reserve of timber was uppermost in the minds of policy makers during the Forestry Commission's first 40 years. And, as mentioned previously, it was the Commission that acquired our wood a few years after the war and after all the good timber had been cut. What we inherited is largely the fruits of the post-war rehabilitation effort. The substantial success and resulting forest of pine and beech is not the Commission's only hallmark: an apple core was left behind too.

In the very middle of the wood, at the cross roads next to the water tank, is a solitary apple tree. It bears fruit in most years yielding angular almost ridged apples reminiscent of that tropical oddity, the starfruit. On picking, the flesh is astonishingly crisp, slightly sharp to taste and lacking sweetness, more I suspect from lack of sunlight than genes. The apples ripen in the middle fortnight of September. When ripe they are very pale green, with a yellowish or sometimes pinkish buff, but without streaks of red. The apples do not store at all well, quickly becoming soft though without developing a nauseous cotton wool consistency. We've been told that these are codling apples and are grateful for the core thrown away, or perhaps even planted, in the spring of 1957 as another lunch break drew to a close and hacking and digging began again as the gang resumed planting the 25 000 or so pine and beech seedlings. The lone apple tree symbolises forestry. What one person plants another is thankful for.

Jane Austen appreciated this. Anxious to tell Cassandra all that was going on at home, at the rectory in Steventon, she reports: 'Our improvements have advanced very well: the bank along the elm walk is sloped down for the reception of thorns and lilacs; and it is settled that the other side of the path is to continue turf'd and planted with beech, ash and larch.' One senses the Austen family at debate, of finally choosing which trees to plant and of considering what would most enhance the rectory grounds for future generations to enjoy. It's the same in our wood: we enjoy what the previous owner chose to plant, including the solitary codling apple.

The wood is now our responsibility. How we care for it, a later generation will discover. Like the baton in a relay race, stewardship of a woodland is passed to an owner for a time. There is work to do, and a running to be done, before another takes over in due course. The aim, like the relay racer, is to pass the wood on in the best possible condition and with the best possible chance for future success.

4

Ruth's gloves

For us, it is just under half-an-hour's drive to the wood; for John it is a little bit less. The route through the north Hampshire downs is a reward of rural, almost rustic, peacefulness marred only by going under the M3 where the lane from Axford approaches the Wheatsheaf Inn at North Waltham. There are two other small villages on the way, one of which is Dummer, where the greatest open-air preacher of the eighteenth century, George Whitefield, held his first curacy, and now is famous for 'Fergie', Duchess of York. I can claim no other link with the Duchess apart from the meaningless one of coincidence: as she was courted by a prince, John and I were falling for a wood.

Fifteen miles seems about right. Anything less and the wood would absorb every spare hour to the neglect of wife and boys and the coining of the equivalent of a 'golfing widow'—a wood widow? Suggestive, perhaps, of a graceful Narnian dryad, but it sounds too spooky and spidery and Margaret doesn't want to be one whatever it is. If the wood was more than about half-an-hour's drive away, such as the much greater distance to the Welsh wood we had looked at, it would rule out putting in a worthwhile morning's or afternoon's work without taking up the rest of the day as well.

On the day after we became owners I began what became a longer job than expected; in fact, on and off, it took the rest of the first year. Really there were two related jobs. One was to mark clearly all the best beech, and occasionally sycamore trees, to ensure they were retained and favoured later on, rather like identifying high fliers likely to make it to the top. The other job was to slash mark for early cutting out the poorest unwanted trees, along with woody undergrowth like hazel. For once this work did not conform to experience. Over the years I've found that I achieve about two-thirds of what I set out to do in a particular time. This rule seems to hold reasonably well whether it's work in the next hour, day, week, or month. Of course, it is only an average, but I suppose I'm an optimist. Setting out to do a lot, and accomplishing at least two-thirds, gets quite a bit done. This ratio helps when trying to plan realistically. It complements that other lesson of experience about who to ask when a job needs doing: go to the person already hard pressed. That person has got lots to do because they do a lot and are willing, within reason, to squeeze in the extra task. The opposite, of course, is what wise King Solomon observed: *laziness brings on deep sleep* (the lazy person gets lazier) and *a little folding of the hands to rest—and poverty will come on you like a bandit*. There was no resting in sorting out good quality from poor in the wood, but it took much longer than expected.

This first job, which was needed over the whole 18 acres of the wood, was to thin and, rather belatedly, clean the broadleaved trees. These two operations remove unwanted woody growth and make space for the remaining trees to grow better. If this was neglected any longer, while the pine strips would look after themselves, the potential for a fine beech woodland would recede. The very best trees were marked for retaining for many years, poor ones and unwanted scrub blazed for early removal. As marking proceeded, barely enough beech were making the grade of final crop standard, and even these urgently needed freeing from the competition of other trees. This is what thinning does: poor trees are cut, good ones given more space to grow. The low proportion of good quality trees meant that this first thinning would yield only very low grade hardwood—bent, forked, or defective trees fit only for firewood. Fortunately, I had recently met a woodland contractor, based in Reading, who was keen to expand his activities on the firewood side. Martin, who ran Smallwood Services, visited the wood in May with three friends to look at the possibilities. There was something like 200 tons of wood to cut, including a few of the poorest pine, though by the end of the thinning Martin reckoned he got less than this. A lot of hazel had been left, yielding little reward for much labour. Whatever the exact amount was, a firewood thinning over 18 acres represented a jump in the scale of his operations, and it brought the usual hassles of expansion.

There are many ways to thin a woodland. The Forestry Commission, for example, had already systematically cut out one row in three of the pine. This simple approach could easily be used for the beech. It is easy to mark, and extraction of the trees along the row being felled is straightforward, but with only a few good beech trees, such an unselective thinning would have reduced them by a further third. Instead, every tree was carefully inspected: it's the best way, but also the most costly in time. This long job was begun by working steadily up and down the three row-wide strips of beech looking for good trees to favour and mark in a clear way. These beech, and a few sycamore, were being set aside to grow for another 80 years or so, so they had to be the best. The removal of poor trees in the thinning would also deliberately favour these selected ones.

When looking for good timber trees, especially among oak, ash, beech, or sycamore, the key feature is a straight trunk. There is no point in growing a large curved tree; even the navy no longer needs the 'knees and crooks' from great, ancient oaks to build

'men o' war' as it did in the days when they were the wooden walls of England. Curved trunks are difficult to saw efficiently, there is much waste, and the resulting timber is likely to warp. Similarly, forked trees are undesirable as are ones with numerous heavy branches, or with decay or other defects. And, if a tree passes all these tests, it must still show enough vigour to merit favouring. Trees, like people, vary enormously in characteristics; but once a slow grower, it is likely to stay so although the extra space a thinning can give does help a bit. Out of every 25 or so trees in each ten-yard stretch of a three-row strip, usually one or two would make the grade, though not always. This discipline of systematic searching forced me to visit every part of the wood and so discover just how widespread was the impenetrable thorny undergrowth, more often than not bound-up with *Clematis*, and just how many active burrows there were. It was not always easy to spy out the quality of the beech and sycamore.

Marking trees gives one 'forester's neck'. Even an hour squinting at their tops causes a stiff neck. And if, while gazing up at the quality of the crowns, you move from one tree to the next, the woodland floor ensures it's not only a stiff neck you get. It is not sufficient just to look at the bottom of a tree (to make sure your own stays intact) even though it's the valuable part; the whole tree needs to be considered, so selecting the best takes time and care. It's a bit like looking round a second-hand car mart: only a few are decent, but many pretend to be so, hiding their deficiencies. Quite often several good trees occur together and only one, ultimately, should be favoured, and one simply has to choose. With beech, tight groups of two or three straight trees can be successfully grown together to maturity if given space around them. Generally, as well as searching for good individuals, the trees should be reasonably spaced apart. There is continual compromise, and no two foresters will ever entirely agree. When a forester's marking is judged to be poor he can get it 'in the neck': in France foresters have been sacked on the spot for not selecting enough trees, or too many, or not the right ones.

In our wood, where forester and boss are rolled into one, selected trees were initially marked by tying round a polythene band. Countless blue and white Boots carrier bags and the posh green and gold ones of 'Marks and Sparks' were sliced into ribbons. The bags, when unfolded flat, yielded strips which were just long enough. Some 500 or 600 selected trees were tagged with these colourful strips, and instead of tying with a knot, most were held in place by a dextrous press of a Bambi stapler. The marking began in late Spring and, by my going to the wood on several Saturdays, and occasionally a mid-week half day, was finished in September. It was then that Alistair pointed out what was going wrong.

The tiny staples, short and only skin-deep in the bark, were growing into the trees, or rather the tree was occluding them like an ingrowing toenail. These specially selected trees were ingesting a diet of polythene and staples: literally a staple diet, but definitely not nutritious even if trees could absorb nourishment through the bark! Jonathan, our eldest son, gave stalwart help with screwdriver and pliers pulling out the offending morsels of metal and, together, we painted white spots instead. At least we were saved the headache of selecting the trees again. It is surprising, though, that such a tiny and superficial attachment would be absorbed. Truly, 'ware the nails, barbed wire, and other appendages in countless hedgerow trees pressed into service as fence posts.

The selection and marking of trees included the patches of broadleaves which were called low-grade and, by and large, we were pleasantly surprised. Quite a few trees were the equal of the best beech in the pine/beech strips and, once the pine and Douglas fir were felled, all the wood would look much the same: mixed broadleaved woodland with beech the dominant tree.

During this first marking stage, I learned what to do with *Clematis*. Many good trees were being smothered, and the creeper had to be dealt with as part of the marking job. Alistair advised me about the best tool to use. Except for the stoutest of their sinewy stems, nothing beats strong secateurs. Sawing simply frustrates, the climber shredding into long slivers, and hacking with an axe invariably damages the very tree one is trying to help, being the chopping block most readily to hand. *Clematis* must be cut right through since even the smallest strands will regrow. Better even than cutting is to pull the whole climber down, but never look up when doing so! A rain of bits and dirt falls steadily all the while one is tugging and yanking at the stem. A second lesson proved more useful. Pull *Clematis* down in damp weather or just after rain. The branches are more slippery, its grip is weakened, and quite often the whole vegetable tangle simply slides off.

In the dry weather, hard tugging usually yields no more than a shower of dust, specks, and bits for one's pains. Gloves are essential since *Clematis* sap can be a painful irritant. Perhaps this unloved creeper ought to be called 'young man's bane'.

The task of selecting and marking the best trees took most of the first summer. It was not the only job. Since we were keen for Martin to begin cutting firewood, I started to mark what he could cut as soon as the best trees in an area had been singled out and awarded their polythene ribbon. The forester's traditional way is to slash a tree to be cut on both sides of the stem with an axe or billhook to make two blazes. In one sense this is even more fraught than choosing good trees; once blazed the tree is destined to be cut, and when it's gone, it's gone. And this was my first attempt at real thinning. I can remember phoning Martin just after he had started to ask how it looked: were too many trees being blazed or too few; did the stand become too open or was the thinning overly cautious? It's all a matter of experience and this was the first lesson in that best of all classrooms.

Thinning a stand of trees is done several times in the life of a woodland. Most trees are planted between six and nine feet apart, but by the time they are mature, and perhaps towering 100 ft. tall, they are so big that 25 or even 30 feet must separate them. The removal of trees in between is what thinning does. It may be done up to half a dozen times as a stand evolves from large saplings to maturing timber, with about a quarter of the remaining trees cut at each thinning. Like my hair, the stocking get less and less with age—unlike it, the quality gets better and better! If left alone the originally planted trees would sort themselves out naturally, with weaker ones becoming suppressed and dying, and vigorous ones dominating. Unfortunately, it is often not the best and straightest trees which win in such competition, but the super-vigorous, heavily crowned 'wolf' trees as foresters call them, usually of poor quality. Thinning helps nature's own processes and helps the owner because it helps the best trees to grow well, and because some yield is obtained from a woodland.

Martin used the thinnings for firewood, buying the trees from us for £4 per ton. My job was to blaze the trees to be cut, his job to fell, extract, and convert them into short split logs. These he bundled up in string bags, each weighing about 40 lb. (18 kg), and sold direct to garden centres in the autumn. The bags were piled beneath a sign declaring 'wood from responsible management of derelict and neglected woodland'. He felt this would help sales, though he didn't ask what I felt about the choice of adjectives! Delivering to the garden centres, Martin was getting £60–70 per ton which sounds a good profit. But I don't think he made much. Cutting and selling firewood is heavy manual work and even with chainsaws and log splitters a lot of effort goes into turning a standing tree into neatly tied bags of 8-inch split logs attractively presented and stacked next to bales of peat, gnomes, and garden furniture.

The trees cut for firewood were beech, sycamore, birch, and an occasional willow along with the thicker stems of overgrown hazel and a few small pine. All make good firewood, except the pine which spits on burning. The market has increased rapidly in the last 20 years, as concern about fossil fuels has grown, and provides an excellent outlet for selling one's poorest trees to help restore neglected woodland. The home-produced charcoal market is less buoyant owing to the messy business that charcoal burning is. However, it is a scandal that we import so much barbecue charcoal made from hardwoods from eastern North America or from eucalypts in Spain and Portugal: such is the origin of most bags in hardware stores, petrol stations, and garden centres. Britain's forests could easily satisfy the home market and connoisseurs say that British charcoal made traditionally from native oak, beech, or hornbeam is without rival. Buying firewood or charcoal makes sense: it's a renewable fuel and one is not contributing overall to the greenhouse effect. Using firewood can be a risky business, and it is essential that wood bought is really dry and not just to touch. Wood cut one winter should not be burnt until the next, else one wastes most of the heat driving off moisture rather than warming the room. A load of logs dumped on a drive looks impressive, though it is rarely more than half a ton and the merchant should be asked when it was cut. If it's fresh, that is cut in the last three months, barter for a lower price. Start with one-third off since that is the amount of water one is buying along with the solid wood, and it's certainly not wanted.

The overgrown hazel clumps, neglected for 30 or 40 years, consisted of numerous thick stems. Martin did not cut very many, as hazel is difficult to handle and often the thick stems are rotten. Of all woodland types, overmature hazel is the nearest to being useless, and it is not even particularly good for sporting, though one redeeming feature is that the wood makes the best of artists' charcoal. And, as every country lad knows, any clump of hazel will yield at least a few arrows, beautifully straight but safe with the thick pith blunting the tip. Thousands of hectares exist as relics from an age when hazel wands, spars, sticks, and hurdles were the fabric of rural life. Their pliable shoots were cut by coppicing every 7–10 years. Since the war coppicing has been abandoned over large areas, though the resurgence in thatching has led to a revival in parts of Hampshire, Dorset, and East Anglia.

Once the thinning work was under way, another improvement job was begun. It is one which foresters delight in, but every textbook says is uneconomic: high pruning. Pruning forest trees is simple compared with apples and pears or even vines and roses. The one aim is to produce a smooth branch-free lower trunk. This is not to make them look nice, or even to prevent youngsters climbing them, but to grow timber free of knots. Knots occur in wood wherever there is a branch: cut it off, the wound heals over, and all wood subsequently laid down is clear of them. The financial return expected from this work is a better price for pruned sawlogs because of the better quality wood compared with unpruned ones, but one has to wait to the end of the rotation—perhaps 100 years after doing the job!

Only the selected good quality beech and sycamore were high pruned. They were just the right size at about 10–15 cm, or four to six inches, in diameter. Mostly dead or moribund branches were cut off up to a height of about fifteen feet, though later some trees would be pruned up to twenty feet or more. Forks in otherwise good trees were cut out as well.

My first attempts at pruning used a cheap saw bought from a local hardware shop and attached to a broom handle. Something deep within my psyche must insist on economies and first trying to make do with some improvisation—the saving of used Christmas wrapping paper mentality. On only the third visit to do high pruning the saw's plastic handle snapped. The resulting deeply cut finger precipitated not only crimson drops but acquisition of the proper tool. Alistair turned up as I was staunching the flow, quite unable to hide my folly. The new saw is slightly curved and screws securely into a five foot aluminium pole to which one or even two more lengths can be added. At the same time, now fully safety conscious, I bought a bright orange hard hat and started using Ruth's gloves. Each birthday my wife's aunt kindly gives me gardening gloves, not flimsy, genteel ones, but Canadian loggers' type with thick, stiff leather for thumb and index finger. They are used all the time except in wet weather. Unlike the benefit of *Clematis* pulling, pruning in the damp not only cascades drips from the saw-shaken branch, but the amalgam of leather, metal, and moisture greases the pole and slowly oozes, soggy and slippery, onto one's hands. Ever since the little accident, tugging at *Clematis* in the wet, high pruning beech in the dry, and chainsaw work at all times are now done properly mittened. Even the three aluminium pruning poles are gloved together with the ends snugly sheathed by a pair of navy blue British Airways oversocks, not for reasons of safety but to stop grease from the screwthreads staining the upholstery when the poles are carried in the car.

Pruning is another cause of forester's neck. And, like *Clematis* pulling, it's best not to look up too much because sawdust comes down in a fine drizzle. The hard hat repaid its investment from time to time, though rarely were heavy, dangerous branches cut off. Indeed pruning branches thicker than 2 inches leaves wounds too big to callus over quickly and more likely to let in decay. Most branches, as they are cut through, bend, crack and then fall owing to their weight. But, sometimes, owing to unequal growth or because they are part of a narrow fork, each cut of the saw tightens rather than easing the pinch on the blade. More than once, prunee and pruner were both gripping, the branch clenching the saw blade and I attempting to tug it free. A really hard tug worsens the problem as the saw bites deeper and the squeeze tightens. A sharp upward push on the pole often effects release and waggling it from side to side can help. Of course, all such problems are avoided by adopting the sound practice of first cutting through a branch about two feet out from the trunk followed by cutting the stump off. But usually I didn't, and other means had to be sought when the long, shiny, grey pruning pole, with saw blade pinched in the cut, dangled like some elongated jaguar's tail out of the space age. By this stage of failing to free the saw, recourse was made to finding a long pole to prise away the offending branch. Only twice did even this ploy fail. And, happily, on both occasions Jonathan was in the wood and we lassoed the branch to pull it down while the saw was worked loose.

Most of the selected trees have been high pruned to about fifteen feet. It is good winter work, offering free exercise and free warmth while all the time improving the stand of trees. Probably one final pruning will be done to lift the length of clear stem to

twenty feet or a little higher if cutting the next branch looks worthwhile and I can reach it. The pine were not worth bothering with, being too big and much too close to the time when they will be felled. All track and rideside trees have been high pruned to let in light and air to ground level, to help the track stay dry, and Alistair's pheasants to sun themselves.

We hope the work begun gives reasonable prospects for a crop of beech trees to grow on after the pines are felled. The tasks have been mostly straightforward, though having to cope with so much *Clematis* was unexpected, and what to do with over-mature clumps of hazel remains an open question. A great deal was still left after Martin had finished. It would be good to resuscitate some of the denser patches and augment Hampshire's future supplies. One of our new neighbours may just be the person to help.

5

Encounters of the rural kind

I grew up a surburbanite with strong rural connections. By the teenage years, home had become 'sarf ease Lunnon' with the 'f' almost aspirate. Herds of 'How now brown cows?' sorted out the vowels—though this oft repeated question was never answered—and research and teaching in the Third World, of necessity, added the consonants and completed the re-instatement of 'south-east London'. This was hardly preparation, however, for encounters in refined rural Hampshire—at least, not for some of them.

The sporting tenant would shoot in the wood two or three times each winter. My first and only meeting with him was at the entrance just inside the gate. Normally the keeper would phone with the dates for shooting and I would avoid going to the wood on such days. On this particular Saturday, unaware of a shoot, I had parked outside the gate and was about 80 yards away, high pruning the selected beech. It happened quite soon after the purchase and I was still finding my owner's metaphorical feet, as it were. I was in rough working clothes and with hair wet with sweat. Feeling hardly presentable, it was not the occasion I would have chosen for this encounter. Andrew, the sporting tenant, surrounded by others arriving for the morning's shoot, watched as I made my way up the track to the gate to discover what was going on. Being alone and conspicuous made one self-conscious. Faintly embarrassed, ill at ease in appearance, and gazing up (again metaphorically) from my 5 feet $7\frac{1}{2}$ inches to one perched in a Range Rover, I asked what he wanted. He presumed I was the owner, explained the morning's intent, apologised for the obvious lack of notice, and asked permission to drive through the wood and along the track by the railway. The meeting was entirely cordial and the confusion was cleared up, though the illogical embarrassment remains the keen recollection. It does not rival the occasion when, at my sister's wedding, and as an even more self-conscious 16-year-old, I announced loudly to elderly and deaf aunts of the new in-laws that 'I was the bride's sister'. First encounters and how one handles them can be very revealing.

Only twice since that early meeting have sporting and forestry work directly conflicted. Once, when with the generous help of our nearest neighbours, the Armstrongs, we were cutting up a fallen pine blocking the main track, beaters turned up to harry the pheasants into flight, which doubtless had already run for cover at the sound of the chainsaw. The other, and more recent, occasion involved my son, who had gone to cut firewood and not only had difficulty in convincing the shooting party that he was the owner's son, but also had to wait until they had finished. Such conflicts of interest have been rare, and never arose in Alistair's time as keeper.

During almost every visit to the wood in the first year or two, I would bump into Alistair. Bump into is the wrong verb, so is encounter or meet: be surprised by is perhaps closest. He had a knack of simply appearing, making me jump at times, not because of his gun slung low at his side, but the fright of becoming suddenly aware of another's presence when deeply engrossed. I don't think he actually crept up on me, as after all the wood is noisy enough with sticks snapping underfoot or vegetation rustling as one passes, but he did seem to materialise out of nowhere. Grunts would be exchanged, followed by the Englishman's predilection, followed in turn by comment on what I was doing. In this way emerged a little about the wood's history; about the time the fire brigade cut the chain at the gate racing to the railway, only to discover they had accessed the wrong part; about Alistair's opinion of local forestry contractors; and about the solitary life of a gamekeeper and the attendant, often uncertain, contact with agents and owners. He has even pointed out a mistake in a booklet I wrote on coppice. Initially, I would ask innocent questions like was there foxhunting? No, who would risk hounds milling around, let alone chasing, right next to a mainline railway. More commonly, my countryside ignorance was revealed concerning field sports and keepering despite 25 years of professional forestry. I greatly valued these encounters.

Alistair's, and the sporting tenant's, only real interest in the wood were the pheasants. During winter—the season runs from 1 October to 1 February—three or four shoots would be held and a few brace taken. Alistair considered the shooting in the wood to be modest, but still worth investing his time to put down feedcorn in a couple of places. This he did by mixing it with straw so the pheasants wouldn't find it all at once. Later keepers used custom-made dispensers or inverted 44-gallon drums with pouches cut out of the side for the grain to attract and fatten their quarry continuously all season and, regrettably, to do the same to unwanted grey squirrels. Pheasant chicks are not reared in pens in the wood, which would have been stretching the terms of the lease somewhat, but everything else possible is done to provide the warm sheltered habitat they seek.

Pheasants are lazy birds, and to provide sport they need winkling out of the cosy undergrowth into rising flight over the assembled guns. Innumerable problems confront the sportsman, not to mention the pheasants. The wood is a bit dense with trees and undergrowth, and once the beaters scare the overfat birds to flight they may corkscrew upwards between the trees and break canopy, just about exhausted, just about out of sight, and just about useless for real sport. The tracks and rides are narrow, too, and mostly shaded which keeps them cool and short of the ideal of sunlit glades beloved of wildlife, hunted or not. There is, of course, plenty of shelter in the wood, where the pheasants congregate as warm summer fields and feeding on grain wane to autumn gales, early frosts, and scarcity of food (hence the corn the keeper dispenses). What the beaters do, strung out across the wood, is to work steadily downwards through the trees to provoke the pheasants into flight over the young plantation at the bottom. This is the flushing cover needed to launch the lazy birds over the railway, and so present a rising target for the assembled shoot beyond.

I've not joined a shoot nor, for that matter, a hunt and don't have particularly strong feelings about them. Preserving the sporting value of the countryside—hedges, spinneys, and copses, and the variety of crops brought such as mustard and kale, and the planting of berry-bearing shrubs like sloe and cotoneaster—are good for both wildlife and landscape. And, killing some animals for meat or pest control is necessary. It is hard, though for one not involved, to see much sport in rearing huge numbers of birds and stocking woods to enormous densities just so that a goodly number can be brought down by even the most inexpert gun. Perhaps as a reason for being in the countryside, it is reason enough. But enough of unsolicited opinions, a temptation to which we are all prey. The antidote I heard one Sunday at the International Church in Addis Ababa. The preacher, with a hint of humour and twinkle in his eye, began his sermon musing: 'When you come to think about it, there are just *two* sorts of problems in the world.' He paused, as his hearers' minds flashed nuclear war, racial hatred, evil towards defenceless children, or blasphemy, and continued '. . . yours and mine.' The ensuing laughter, responding to deftness of touch, underscored the pervasive truth of selfishness. Likewise G.K. Chesterton owned up for us all with his immortal reply to the question in the *Times*: 'What's wrong with the world?' 'I am', he replied. I shall leave country sports to those who enjoy them.

Since Alistair left, another keeper has come and gone whom I never got to know well. The first encounter with him arose when Stephen, our middle son, found a superb U.S. army knife lying in the road. We had been camping at the wood and were on the 3-mile lunch walk to treat ourselves at the Little Chef at Popham when he spotted it. He picked it up. It was new, razor sharp, impressively solid and shiny, and every young scout's dream. A little wistfully, he balanced it gingerly on a prominent post beside the road. The postman was delivering letters to a nearby cottage—the keeper's as it turned out—and he passed on the news of the finding to the new keeper, John, who had replaced Alistair, and whom we had not yet met. Later that day, as we were returning with at least the boys full of chips, we met both postman and keeper in Waltham lane and introductions were effected. John, the new keeper, was suitably grateful for the discovery and return of the splendid knife. He was surprised, though, to learn of our camping in the wood for a couple of nights, having seen neither car nor tent, and betraying his infrequency of visits compared with Alistair. His phone calls were equally infrequent and it was usually necessary to coax out of him the dates of each season's shoots. The knife episode was almost the only encounter we had, but not quite, as the next incident relates.

For all the time we had known the wood, it had stood in the middle, a resolute sentinel at the wood's cross rides. One Saturday it vanished. Squinting between rows of trees it could be used to orientate oneself, but not on this day. I was at a loss. A few more strides, hurried now by uncertainty, returned me to the middle ride. Sure enough, it was gone. It had been cleanly taken: the large grey water tank. All that was left behind was a brown rectangular hallmark tracing its dimensions, the two railway sleepers on which it had stood, and the dank scrapings of 30 years. These last advertised their whereabouts to the nose as they festered in the hot June sun. This was no common theft, a galvanised tank of 6 foot by 4 foot by 3 foot isn't an easy picking, it wasn't visible from the road, and the gate hadn't been broken nor lock and chain cut through. The culprit, as discovered on the third phone call that evening, was the new keeper who simply saw the tank as suitable for another job elsewhere, and had not appreciated it had been specially placed centrally in the wood in case of fire. It was quickly restored to guard duty next to the cross rides.

The estate of which our leasehold wood is a part has an under-keeper as well. Encounters are so infrequent that introductions are needed almost each time we meet. The present headkeeper I've

not met beyond the other end of the phone and a new padlock appearing on the gate chain.

The lessor, when we bought the wood, was the Prudential Insurance Company. The past tense reveals the fact of two changes of landlord since then. From the wood the Pru and later landlords received income from letting the sporting rights and our peppercorn rent of £2.72 per year. This ground rent is for the whole wood, it is not even per acre, and derives from the 2/6 (two shillings and sixpence) per acre agreed between the original owner and the Forestry Commission in 1953. And it was fixed for 999 years. Apart from this annual transaction which John helpfully looks after, the only contact with the Pru concerned the big trees at the bottom beside the railway. These belonged to them, an asset they decided to realise in 1987. We discussed with their agent, Mr Humphris, questions of timing, access, restitution of the track, and what to do with the patch of sycamore coppice once they decided to fell this timber. Once felled the land became ours.

The Pru had acquired the Steventon Estate, at least the part that included our freehold, in 1976. At the time, the wood was leased by the Forestry Commission, but this arrangement had not always prevailed. Encounters with the past through old correspondence, notes, and from enquiries provoked curiosity as to why, after the war, the wood changed hands into public ownership.

In 1951 the wood and several others on the Steventon estate were described as 'devastated'. This term, long since fallen into disuse, was applied to woods which had done their bit for the war effort—cut rather than dug for victory—and then left to their own devices. A tangle of dense undergrowth and impenetrable thicket was always the result. Although today it would perhaps be cleaned and encouraged to develop its potential, in those optimistic post-war years of order, control, and imposition upon nature, epitomised by the East African groundnuts fiasco, coniferisation or replanting with close-knit rows of beech or oak, often in mixture with pine or larch, were standard prescriptions. Forestry Commission researchers had conducted experiments into 'rehabilitation of derelict woodland', which compared different ways of coping with such thicket and approaches to replanting. Clearing, cutting swathes, and enriching natural gaps, using swipes, mowers, and massive rollers, and other jungle-busting kit in the days before herbicides were all tested across half a dozen sites in southern England. It was all very expensive. When I reviewed this

research 30 years later, the 'do nothing' treatment, where the previous generation of scientists had had the good sense to leave part of each derelict woodland experiment wholly untouched, had sometimes turned into reasonable stands with a scattering of nice timber trees. But it was not always so.

The Steventon estate's devastated woodlands were causing concern. As well as being of uncertain silvicultural promise, neighbouring farmers were upset by the hordes of rabbits harboured there. The Hampshire Agricultural Executive Committee of the National Agricultural Advisory Service were anxious to do something about this infestation. They were keen for the Forestry Commission to acquire the woods primarily because they knew that vermin would be controlled. Indeed, I recall our forestry lecturer at Bangor, Tom Owen, telling the class of '68 that one could always spot a Commission wood by its stout, well maintained fence. The owner of the estate appeared not particularly keen to replant after the war, his primary interest being the sporting, and there was no requirement that he do so.

Negotiations for acquisition by the Forestry Commission began in late 1951. The key terms were agreed in the following spring with completion a year later. Forestry Commission director, O.J. Sangar, was the final authorising signature which acquired for the state the right to the land for forestry purposes for 999 years. The owner would be paid half a crown per acre per year and would retain, principally, the sporting rights. Also retained, but only for a further 50 years, were small areas of young mixed conifer and hardwood which constituted the mature timber at the bottom of the wood when we bought it. There was even rent abatement for the lessee for the land with the reserved timber and other provisions which the Forestry Commission acquired, and we inherited. These included permission to build a forester's cottage without undue hindrance—from the lessor if not from the district council—and the right to control vermin and shoot deer, along with the lessor. That last ambiguity is even worse than the motorway warning 'Roadworks, delays possible until July 1996'! By the time we acquired the wood not many years remained before expiry for the reserved timber.

The Prudential sold the reserved timber in 1986 and the logs, each weighing up to 2 tons, were dragged 200 yards up our main track and stacked at the entrance. Just enough room was left for a lorry to load and to turn. As Martin finished taking his bags of firewood, timber-sized logs took their place. Despite working in February, the track did not rut too badly, and the Pru were responsible for its restitution in any case. The job of felling and extraction went to a local contractor, Henry Robinson and Son of Deane. At the entrance the logs were divided into different piles of larch, of oak, and of other conifers and hardwoods. They went in dribs and drabs throughout the spring, which is always a difficult time to market timber since mills are cluttered as owners, finished for the time being with sporting, want to get thinning and felling done before pheasant rearing begins in earnest in early summer.

The main track was not restored immediately despite a couple of reminders to the Pru from me and one from John when he sent off the payment for the annual rent. In mid-June, and beginning to wonder if the job would ever get done, I was at the wood with my mother, when two strangers appeared. Two and a bit years of ownership had lengthened the stride and firmed the voice and I asked what they were doing. They had come to survey the track, and this on a Saturday afternoon, to estimate the cost of regrading. This they did a fortnight later: car access to the bottom of the wood again became assured in all seasons, and another local firm was noted down for future reference. This use of local contacts is important for keeping rural industry and life in the countryside intact, quite apart from being the best people to employ anyway, being familiar with the conditions and with a reputation to maintain.

This chance meeting was the second with strangers in the space of a few weeks. Earlier an unmarked 1½ ton van was parked at the bottom of the main track next to the railway. Two men sat inside munching sandwiches and digesting the *Sun* and *Mirror* propped against the steering wheel and glove compartment respectively. They were British Rail staff intent on inspecting the line, or the grey transformer, but obviously not until after lunch. It was a nice spot they had found. I learned that BR had recently discontinued using their familiar yellow vans and were leasing vehicles as an economy. British Rail's right of access goes back many years and they appear to have no responsibility toward track maintenance—ours not the main line—despite being the most frequent user. British Rail must be able to reach their railway lines throughout their length and, as mentioned in Chapter 3, the confluence of level ground and railway track for the length of the wood makes it a key location. BR have contributed two padlocks to the gate chain making, until recently, three in all. They have let me have a key to one of theirs ever since they once left the gate locked with our padlock excluded from the chain links, dangling helpless and useless.

Encounters with officialdom in connection with the wood have proved very infrequent. The local Forestry Commission office issues felling licences and awards planting and management grants, but I've only needed to contact the forester responsible for our part of Hampshire, Robin Hendrie, a couple of times at most. I expect he has made a few visits over the years, before sanctioning fellings for example, but it's usually not necessary to disturb the owner for this. Hampshire's woodland officer, Jonathan Howe, who has done so much to advance the county's constructive woodlands policy, has visited once. He was unimpressed by what he saw of our scraps of old hazel coppice, which was the purpose of his visit. He did see with interest, however, what I had not consciously noted, that remants of hedgerow trees, shrubs, and woodland flowers still delineated the wood's four original field boundaries. As for others, unlike farmers, we are spared vets and their fees, and salesmen and their visits, and milk lorries needn't call every few days or the postman to bring ever more forms and returns. Few officials, or anyone else, need make their way along the lane to our wood for reasons of business.

At the cross roads halfway between North Waltham and Overton one turns left to get to the wood. The lane runs along the ridge and has passing places carved out by innumerable vehicles mounting the bank. One bright, and not especially chilly, November morning in 1989 this final stretch of lane was blocked by a broken-down car, its orange winkers anxiously flashing, some 20 yards north of the corner of the wood. The lady said someone had gone for help. No amount of hopeful peering under the bonnet on my part would cure the problem, and soon Nick arrived. He turned out to be our nearest neighbour, living a couple of hundred yards beyond the far end of the wood. The lady was doing the school run for Nick and Sally's children. The rope from our boot was used to tow the car back to Nick and Sally's house, and thus acquaintance was made. They can see our wood from their home and, of course, frequent this lane that runs along the top. Nick has bought a little firewood, but much more importantly the Armstrongs have kindly reported damage to the gate, helped cut a fallen pine blocking the main track, were in touch following the storm of January 1990, and alerted us to local land deals. Their kindness and help, including tea we've enjoyed, is as lovely as the peacocks which strut around the garden and stand parade on their garage roof.

At the same cross roads, on a triangle of grass, a yellow gypsy caravan with cob, cart, and all arrived one late summer's day. Melvyn, the owner, works hazel coppice fashioning spars for thatching and travels the lanes of southern England from Chichester to Glastonbury following his work. North-east Hampshire and, indeed, the grass triangle, have become a favourite and he has settled there for several autumns before going to lower ground for the winter. His caravan is Yorkshire-made dating back to 1907 and is a familiar sight around the lanes. A yellow frame hoop bears the bright emerald canvas canopy on top of the equally bright yellow wooden cart borne up by leaf springs on the large spoked wheels. A door opens out to steps that lead down between the shafts for the cob. The spars Melvyn makes, just like those Marty endlessly fashioned in the small hours in Hardy's *Woodlanders*, come from young hazel, which is now in such short supply. Melvyn has looked through the wood, since it's close to hand, to see whether any of the overgrown clumps can augment his needs. Old hazel will often have a few shoots growing up through its dying centre, but it cannot be relied upon. Melvyn will soon become a more permanent neighbour, buying part of the field on our south side to grow thatching straw, which is in similarly short supply. Indeed, Britain imports both thatching straw from Holland and hazel spars from France, and there is even a threat of plastic spars displacing the genuine article. Between Melvyn's land and our boundary Mike and Annie from Overton are turning the strip of field into a market garden dedicated to organically grown vegetables. We meet almost every time I'm at the wood and more helpful neighbours would be hard to imagine. They come into this story a little more in later chapters.

I've not met our neighbour who farms to the north of the wood (our boundaries don't in fact quite meet), though we have often been startled by his bird-scarer. He places this gas-driven contraption just at the edge of the small wood that is adjacent to ours, and its boom reverberates all over, frightening pigeon and people alike. Two loud bangs repeat in quick succession, followed after a moment or so by a third that coughs a whimper, and then silence for an indeterminate time before the scarer rouses itself to startle again. It must be powerful since car horns blasted at the entrance gate to attract attention cannot be heard farther in than the grey water tank in the middle. The trees muffle sound well. From the bottom of the wood, only British Rail's oldest rolling stock are noisy enough to be heard at the gate; newer trains are very much quieter. And they don't startle, announcing their arrival a minute or two before with echoes from Litchfield tunnel or distant rumbling as they take the bend high on the embankment overlooking Steventon.

Never did we think that buying a wood would also bring so many encounters of the rural kind from organic farmer to hazel coppice craftsman, and from acquaintance to neighbour and friend. All such people were real and rural. It was up to us to complete the evolution from suburbanite with rural connections to local small woodland owner. We declared our credentials and commitment to the long term by doing the one thing expected of foresters: plant trees to create new woodland. The Pru's early felling of the big timber at the bottom provided the opportunity in this way to make our pledge to the well-being of the countryside.

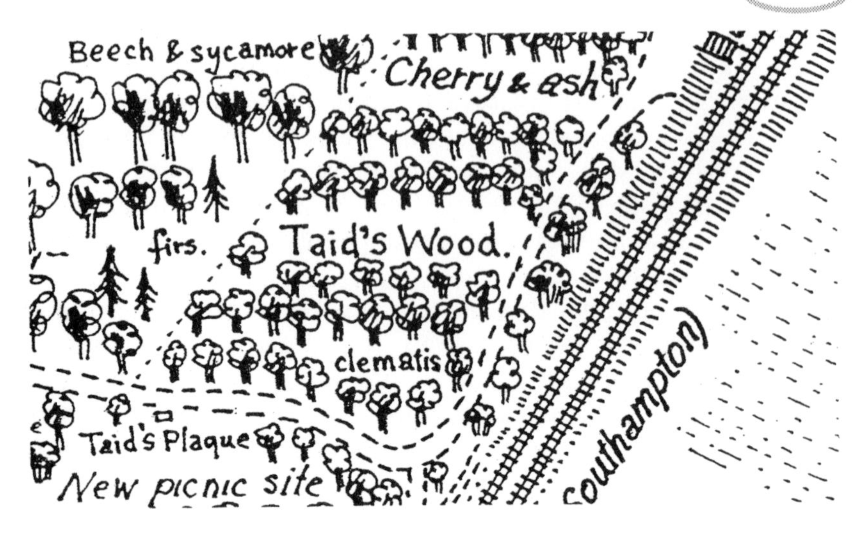

The planting of Taid's wood

As already mentioned, when we bought the wood in April 1985, 1.6 hectares, or nearly 4 acres, beside the railway were not ours to use. On this land the mix of larches and other conifers, sycamore, and oak trees had not been felled during the war because it was younger than the rest of the original planting. At least, so we surmised, concluding that the trees had probably been planted around 1910 rather than 1895, our estimate for the rest of the wood.

At the time the Forestry Commission acquired the wood and two other small ones from the Steventon estate in the early 1950s, each had a reserved area of older timber, though ours had the biggest. The lease stated that these trees need not be felled until 2003. The Prudential decided not to wait and sold the trees to Mendip Forestry Ltd in 1986, who in turn employed the local firm of father and son, Henry and Alan Robinson of Deane, as contractors. Felling began in late autumn and each log was dragged muddily up the main ride and stacked by the gate to await collection. The Pru had sought permission to do this. By March all the old trees at the bottom of the wood were felled and extracted, and the 4 acres cleared and tidy with all but the smallest branches and other debris piled up and burnt. The ashen white remains of the fires peppered the gentle slope and the now uncovered ground revealed even the smallest humps, hollows, and rabbit burrows. I have rarely seen such a cleanly prepared site.

John and I were delighted to acquire the bottom 4 acres 16 years ahead of time. We had had notice of this and set aside the last week in March to do the replanting, a total of some 1500 trees. This new area was named 'Taid's wood' after my father (Taid is Welsh for grandfather) who had died in June the previous year. My mother spread his ashes among the well-grown sycamore coppice which divided the planting into two, and on the anniversary of what would have been his eighty-first birthday the family met at the wood to erect a plaque commemorating his 80 years. My father was a skilled amateur entomologist, learning his trade in the Surrey countryside around Ewhurst beneath Pitch, Holmbury, and Leith hills, the great greensand outcrops rising out of the Weald to achieve the highest elevations in south-east England. He was a careful observer. For years he recorded the moths circling the one bulb British Railways left glowing at Petts Wood station for reporters and journalists returning from Fleet Street in the small hours. His conclusions from the observations for 1947 were published in *The Entomologist*. A few years later 600 tiny privet hawk moth caterpillars came with us on holiday and each day scouring the locality for privet leaves came before the beach, and sometimes even before breakfast. Interest in privet hawk gave way to northern eggar moths as successive generations were bred from a few caterpillars collected on holiday in Rhyd in Snowdonia in 1954. I have his collection arranged in geometrical precision to show their lineage and forms. Some of the progeny, and that of crosses made with the southern oak eggar, now reside deep within the Natural History Museum. The smell of

Oak eggar moth

paradichlor preservative is part of my childhood, and I thought all fathers bred moths and even counted them circling lamps at two in the morning. The wood is a suitable resting place, and I am immensely grateful for inheriting a love of the countryside despite a suburban upbringing.

On the 4 acres we followed British tradition. For in Britain replenishing woodland after it is felled is almost always by planting, or sometimes by coppicing in broadleaved woods if only sticks, stakes, and poles are wanted. This predilection for planting is peculiarly British. As long ago as John Evelyn, writing his *Sylva* for the Royal Society in 1662, emphasis was on replanting. Perhaps this betrays us as reluctant Europeans, for on the Continent natural regeneration is relied on where the new crop of trees is carefully nurtured from seed that falls from the old. This is quite possible in Britain, but patience is needed to wait for fruitful seed years 5 or even 10 years apart in the case of oak and beech. Also the rising populations of rabbits, burgeoning numbers of deer, and Finlay Macrae's maggots of the hillside and other not so woolly livestock, all 'spoil the conception' as Evelyn would have said. Even birds, mice, and grey squirrels add to the depredations, the last excising surgically the growing point of unfolding cotyledons in late spring. In these hurried times we simply don't wait. As Harry Frew, the forester at Clarendon Park, would observe, 'Informed inactivity may be the mother of inebriation, but it is the father of forests.'

We did not seriously consider natural regeneration for Taid's wood. The previous crop was of indifferent quality and not worth perpetuating, and I was keen to create an entirely broadleaved woodland. Having written the Forestry Commission's standard reference on how to grow such trees, professionally I really had no choice!

Choosing which species of tree to plant is fraught with difficulty. The decision was needed many months in advance of planting both to apply for a grant under the then Broadleaved Woodland Grant Scheme to help defray costs, and to order the right seedlings from a nursery. News had come in early summer of 1986 that the land would become available the next spring, and thus we began to ponder the choice. Having ruled out natural regeneration and caring for the mix of birch, sallow, sycamore, and the odd oak and conifer which would have arisen, we turned to consider the site's potential. But the real reason the decision is so fraught is that once made it is fixed for the lifetime of the crop.

For the next 50 or 100 years, or even longer, the forester's skill is displayed for all to observe, in much the same way buildings do for their architects. Unlike the farmer's annual harvest and equally annual opportunity to reconsider, the forester at best has a once-in-a-lifetime opportunity. Largely for this reason past generations of tree planters put in the 'planter's mix', some four or five different types of tree with the intention of selecting the best one or two to grow to maturity as the young stand sorts itself out.

Taid's wood is beside the railway and is well sheltered due to its mid-slope position. This is helped by the tall beech and Douglas fir trees of the older wood which fringe the upper western edge. Also a wedge of sycamore coppice, where in fact Taid's plaque is, bisects the ground into larger and smaller parts and further adds shelter. Most broadleaved trees, except birch, rowan, and sallow need moderate shelter in their early years for good growth and straight stems. Indeed, too much exposure is one of the reasons why broadleaves are not extensively planted as a timber crop in the uplands.

The gentle slope faces north-east, in the direction of North Waltham village and Basingstoke in the distance. This is the best aspect for broadleaved trees (the compass point, not Basingstoke) because in early summer, when new growth flourishes, temperatures are coolest and soil moisture levels kept up. Frost damage to newly planted trees can be a hazard on east facing slopes, but aspect is a small consideration compared with avoiding frost pockets. The key is either not to plant sites where cold air collects on still, starlit nights in spring or to encourage it to 'drain' away, in the same way as in design of orchards.

The soil is quite deeply rootable: a clay loam of 9–12 in. overlying chalk rubble. Free lime in the form of chalk fragments, along with bits of flint, litter the surface making the soil alkaline. This constrains what we could hope to grow in the same way such soil prevents a gardener cultivating rhododendrons, azaleas, or heathers. Most conifers are ruled out, Corsican pine, yew, and red cedar excepted, and a few broadleaves such as sweet chestnut and common lime. Broadleaves were decided upon, but not exotics such as southern beech, eucalypts, and poplars for reasons of market uncertainty despite their demonstrably superior vigour, and my prejudice.

The final decision was taken over dinner with John and his fiancée, Gill, after the alternatives were laid out rather more congenially than in the above analysis. John's remark that he wanted to get something back in his lifetime was the decider. Assuming he enjoys at least the Bible's allotted three score and ten plus another ten thanks to modern medicine, that added up to about 45 years. For native broadleaves this is not very long. With the right conditions and good care ash will make small timber size in 40–50 years and there is a possibility that, even earlier, the very best stems might produce valuable wood suitable for sports equipment. Ash was chosen for 70 per cent of the new planting, with wild cherry and oak a further 15 per cent each. In this way the ash provides early revenue, followed by cherry, and lastly by the oak as a final crop which our great grandchildren will appreciate.

Ash is an excellent timber, much in demand, and it is a good tree for chalky soils. William Cobbett, when passing through Micheldever forest, a few miles to the south of our wood, in 1822 specially commended the species for Hampshire's chalk downs. Textbooks insist that it requires fertile soil, and this comment is included in the one I wrote for the Forestry Commission. The ground in Taid's wood doesn't quite make the top grade although Dog's mercury carpets everywhere and is a useful indicator plant of sites good enough for ash. The ingredients for good ash growth are control of competing weeds when trees are young, and plenty of nitrogen from rich organic matter or mineral fertiliser once they are well away. At about this time I was relating to Mike Jarvis, a friend from church, how we chose what species to plant and he remonstrated that surely a proper soil survey was essential for such an important and incredibly long-term decision? Mike, the author of several manuals which map and describe the soils of southern England was, of course, quite right. My excuse for not digging several soil pits was that plenty of rabbits and not a few moles had burrowed holes and built hills, so revealing at irregular intervals the essence of numerous soil profiles.

The trees ordered for the planting were small, either one or two years old and none taller than 18 in., i.e. 45 cm in the nursery trade catalogue. Apart from being much cheaper, smaller ones establish themselves readily and soon grow with vigour. Often within a few years after planting small young trees catch up larger ones put in at the same time, as the latter struggle to get a hold. The bigger crowns of large trees die back, usually because of inadequate roots with little time to get established: they can't cope with the shock of transplanting. This lesson is only now being learned as the legacy of imaginative planting designs on paper, from the pens of optimistic landscape architects seeking instant effect, litters sub-urban and industrial developments with half-dead trees. Even Evelyn, writing in 1662, appreciated this point, depicting Leviathan-like contraptions for transplanting semi-mature trees, all with the object of excavating and transporting the largest poss-ible rootball. Even then, like moving house, it takes time to settle and to put down new roots! For a time this means thin crowns and poor growth. For planting trees, Schumacher's aphorism applies: 'Small is beautiful.'

The young ash and cherry trees were delivered from Oakovers nursery in Kent in late January. Some days later the mix of English and sessile oak transplants I had bought were collected from the Forestry Commission's research nursery at Headley, near Alice Holt Forest. Our small car was loaded to the roof by this forest in waiting. The Talbot Horizon had been my father's, and he would have thoroughly approved of such a use. The plants were bare-rooted and in thick polythene bags each containing 5–10 bundles of 25 each. They were stored, still sealed in their bags, in a cool dark corner of the garden shed. Provided they do not get over-heated or frosted, it is safe to keep bare-rooted tree seedlings or transplants like this for up to a couple of months.

One other important decision remained. How to protect the young trees from the attentions of rabbits and deer? Without pro-tection all the trees would be browsed; very few parts of Britain escape such a hazard. Although there are one or two exotic unguents to smear on trees as repellents—most that are tested including the much-hyped essence of lion dung fail to work as hoped—there are really only three ways to stop browsing. One can exclude animals from the whole area by fencing, individually protect each tree using a treeshelter or guard, or eradicate the offenders. The last was impractical, as even a 90 per cent cull of rabbits has no effect owing to their not-so-metaphorical reproduc-tive reputation. Fencing typically costs £2–4 per yard (metre) to achieve a standard sufficient to exclude roe deer. For our elon-gated rectangular planting areas this would be inefficient: a lot of fence enclosing only 1.3 hectares. We plumped for individual tree protection using treeshelters.

Treeshelters, widely known as 'Tuley Tubes', are the invention of Graham Tuley, a fellow Forestry Commission researcher in the late 1970s. Graham surprised upon the treeshelter idea in his search for an idiot-proof way of planting, protecting, then leaving trees to get on with growing. We rather eschewed his ideas in 1979 when his experiment of 80 young oaks were variously enclosed in plastic netting, some of which were also covered with a sleeve of polythene. The cynics considered that when the sun really got to work in early summer, greenhouse-high temperatures would put paid to the trees. We were wrong. Far from dying, the plastic sheltered oaks grew best, and really astonishingly so in the second year. At the Alice Holt research station one can still see Britain's fastest grown oak which reached the top of a 13-foot-high experimental Tuley tube in just three and a bit years!

Improved growth isn't the only or principal benefit of treeshelters. They protect the tree from all forms of mammal damage, they greatly simplify the careful use of herbicides to kill weeds in the all important 18 in. wide zone around a tree, and most importantly they identify where the newly planted tree is. It is easy to lose a young seedling of ash, oak, or beech amongst the lush herbage of a lowland site; the treeshelter is its beacon. At times they attract too much attention when whole fields of tubes are no better than the vilified 'serried ranks of dark conifers' despite the native broadleaves usually inside.

We used Tubex treeshelters, a rather late entrant into the treeshelter market, but much the most widely used today. They were 4 feet (1.2 m) tall, a pale translucent brown, and round with a splayed top. In mid-March 1500 were delivered direct to the wood where they were hidden in their bundles beneath an unusually dense row of pine about 30 yards in from the gate. Stakes to secure the treeshelter were thin split or sawn chestnut from Homewood's sawmill in Liphook, Hampshire. By the spring equinox all was ready. The plants were in the garden shed, the treeshelters at the top of the wood and their attendant stakes piled at the bottom, and the site beautifully clean. Describing these arrangements in a page or two belies the planning, phoning, and visiting needed to organise it all.

John and I had agreed on the week to take leave to do the planting though other commitments meant that neither of us could be on site every day. Also, it soon became clear that one week was not enough even though we had the help of Douglas Clarke for the first few days. Douglas had just returned from northern Kenya where he had been working at a remote mission station on the north side of the Chalbi desert. I had once visited this outpost at

Kalacha in 1982 on behalf of the Africa Inland Mission to advise on tree planting, and had got to know Douglas through discussions on forestry prior to his going there. Douglas's help now was a god-send. Each morning he bought his long wheel-based Landrover to commute up and down the central track carrying treeshelters to the planting site. It got very muddy, it vibrated, it appeared to consume more oil than petrol, and it exhaled like a dragon, but it saved us much wearisome carrying.

While Doug did the transporting I began the first on-site job of laying out the planting positions for each tree. The mixture of three species (ash, wild cherry, and oak) was organised so that alternate groups of oak and wild cherry occurred at 7-metre (22 ft.) centres—I worked in metres—with ash used as the infiller in between. Individually, trees were planted between 2.5 and 3.5 metres apart or about 8–11 feet. The chestnut stakes were used to mark the planting position and were driven in to await their tender ward and its protective treeshelter. The first row laid out was aligned parallel to the main track and then each subsequent one was offset the correct distance using a stick; the same stick served to space the planting positions within a row. It was a slow task, continually going back and forth to pick up bundles of stakes and then to place them individually in the right spot.

The planting itself involved three operations. First the stakes for the treeshelters were driven into a hole opened up with a crowbar. They needed to be firm and, importantly, less than 4 feet (1.2 m) high so as not to scuff trees once they grew above the shelters. Secondly, the small trees themselves, untied from their bundles, were planted within a few inches of the stake. The job of planting was straightforward. A single swipe drove the mattock into the soil, then it was levered back, and the roots gently inserted into the slit just as far as the root collar and no deeper. A spade would do just as well. Next, while holding the tiny tree absolutely vertical, the brittle, tender roots were quickly and gently teased out, like getting flex ready for wiring a plug, and the soil firmed around them with the ball of the foot. Lastly, for our planting, the treeshelter was placed over the tree and tamped down a little into the surface soil, and then secured to the stake using the black nylon ties with their neat ratchets. The process sounds clinical, but was hard, tiring work for two normally deskbound, and not so youthful, executive types. The weather during the week was damp and fairly cool, ideal for tree planting. Steadily the bags of

trees in the garden shed diminished and the plantation grew. But, by the end of the allotted week's leave, we were far from finished.

John and I shared the jobs, each of us for a time stake-bashing, for a time planting, and for a time shelter-fitting. Each operation drew different blisters on our soft hands, and so we shared those out too. We even shared out the species: John planted most of the ash, and I the oaks and wild cherries. The utter change of work for us both for this week of creating a new woodland was reward in itself. Plodding describes best the rate of progress while beside us morning trains rushed to London and evening ones rushed home. But slowly, day by day, the planted area grew. During one coffee break John remarked on the rustic nature of the work and how satisfying it must be for city commuters travelling the Winchester to Waterloo line to behold yokels labouring in tree planting, just like the peasants we British would observe when travelling to Marseilles for winter warmth or the Alps for skiing. We enjoyed being peasants. We enjoyed too amusing thus the passing traveller who saw not a solicitor and civil servant.

By the end of the week—and the Friday was the gale of 29 March on a day when John laboured alone, sometimes standing, sometimes bending his 6 feet 4 inches, and sometimes avoiding falling trees—the whole site was laid out and the larger of the two areas completely planted. About two days' work remained.

Without Doug's three days we would have only achieved about 60 per cent—a further proof of the two-thirds rule. To finish off I took two extra days' leave during the second week in April to plant the remaining oaks and ash. The cherries had been planted as soon as possible since their buds had been swelling and greening even in March and by mid-April were well into flushing. Oaks typically flush in early May, and ash two or three weeks later. Despite the vagaries of our summer weather, and despite what the proverb might suggest, 'oak before ash' is the norm, even though the concomitant 'splash' rather than 'soak' may not be.

Creating a woodland was satisfying, though it brought a new emotion. From almost the first day I began to be anxious about the trees, each a mere twig in size; it was just like those early weeks two years before, worrying whether 30-year-old trees were old enough to look after themselves! This time the worry was that of the inveterate farmer. Wasn't the weather turning a bit dry, hadn't we had too much rain and, worst of all, would John Kettley's –2°C of frost burn off the newly flushing growth? Within days of the last trees going in on 9 April, the weather turned glorious and temperatures rose to 25°C; just right for speeding bud break and opening of leaves. Indeed, by the end of the second week in May all the trees—ash, oak, and wild cherry—were well flushed. The normally late ash was three weeks early. The species flushes late, avoiding spring frost blackening its soft tender foliage; only walnut is even later and with even more tender leaves. Night after night in May, Michael Fish, still five months away from his gaffe of discounting rumours of the great storm of '87 only hours before it broke, forecast cool unspring-like weather, ground frosts, and air temperatures hovering around zero. Never have I scrutinised the BBC's weather bulletins so closely.

Afflicted with the malady of possession of two years before, excuses were contrived to pop over to the wood to visit the newly planted trees to see if they were all right. This time the frequency didn't diminish with time; if anything, visits increased as the new crop began to surprise. By 20 June the first tree, one of the wild cherries, emerged above its shelter. In the two months since flushing it had grown more than three feet! We had planted, the BBC weather centre had forecasted, but God gave the increase.

In all we planted about 1500 trees over the 1.3 hectares resulting in an average spacing close to the 3 metre (10 ft.) maximum permissible for claiming grant. This spacing is really rather wide for broadleaves, overly limiting the choice of trees for the final crop and not providing enough mutual shelter for good upward growth and suppression of side branches. However, the fewer trees one plants the cheaper it is and the 3 metre rule is a compromise. The total grant received from the Forestry Commission was £920, and the cheque came promptly. Two instalments of £250 would follow after 5 and 10 years. Totalling up all expenses, the initial grant nicely paid for all the materials: the plants, the treeshelters, and the stakes. Our unskilled but committed labour had been given free. Free, too, we hope, has been the pleasure for passengers of seeing a new woodland steadily regenerate season by season and year by year beside an otherwise rather bare stretch of railway line as up-trains emerge from Litchfield tunnel bound for Basingstoke and London beyond.

7

Mother-in-law's dustbin

The Forestry Commission's grant not only helped with the planting, but a small sum was used for improving the patch of sycamore where Taid's plaque is found. Cleaning out of undergrowth, getting rid of some tired hazel, and some early thinning among the coppice shoots were all needed. Provided woodland is under 20 years old, the Forestry Commission helps with the costs of the unremunerative work of rehabilitation. This assistance was introduced in 1985 and is a real help with what has often been dismissed silviculturally as worthless scrub.

The sycamore area only amounted to 0.3 hectares and was partly of coppice origin and partly of seedling growth with the odd oak, birch, and sallow in between. This small distinct copse arose from a fire in about 1968. It divided the two planting areas and it made sense to include it for rehabilitation since it was still young enough for grant aid. Work started in the week before we began the great planting of Taid's wood, though identifying which trees to favour had been done some time before. The best trees or shoots in each clump of coppice were marked and the poorer ones cut out. This treatment, called storing coppice, was one of the ways trees were recruited and grown to timber size to become the standards in coppice with standards.

The coppice stems were quite thin with up to half a dozen springing from each stool. Like clumps of giant rhubarb or celery with wooden stalks some 30 feet tall, they needed thinning out and, while not able to provide food, they were good for fuel. I began the cutting of unwanted stems myself using a small chainsaw, with hard hat and gloves again pressed into service. Although never formally trained, chainsaw skills were learned during research in Swaziland when confronted with 500 pine trees to fell and section so that their annual rings could be measured. While I was operating the chainsaw, and acquiring tingly fingers from an early model without an anti-vibration handle, my wife of a few months was becoming intimate with tree rings—more than a quarter of a million of them we reckon. When asked whether she worked after we got married the reply is unequivocal, though she received but a pittance from my small research grant. Her measurement of the widths of so many annual rings helped to reconstruct the year by year development of the pines, allowing us to compare the growth of the first crop with that of the young second one. This research, and Margaret's key role, has been an enduring blessing. We have returned six times to that jewel of African countries to reassess the progress of the pines. Happily Margaret's ring measuring days never needed repeating and have been replaced by more leisured stays in the Foresters Arms Hotel in the unpronounceable village of Mhlambanyati. But the immensely valuable dataset has allowed comparison so far of three successive crops of pine on the same site. The good result is that each crop or rotation of trees has done a little better than its predecessor.

To thin the sycamore coppice I at first borrowed a small light chainsaw with a 12 inch drawbar from David. He is a church elder and one of those dear people who possess every tool—his car always stands *outside* the garage—and every kindness and generosity to lend and help. The saw had a narrow chain and needed sharpening about once an hour when in continuous use. Its one drawback was a faulty off switch and the choke was used to staunch the noisy two-stroke. All chainsaws, including the one we subsequently bought, need as much attention as a clapped out car advertised in the local rag as 'a good runner'. Hourly chain sharpening with a round file, and equally frequent replenishing of the two-stroke mix and the chain oil, are the minimum required just to keep going. A dirty plug, dirty fuel as the small tank is filled in the most unhygienic of surroundings, and a clogged air filter are commonplace problems. And, starting a two-stroke is a knack: when cold ours normally fires on the fifth pull of the starter cord after pre-determined fiddling with choke and accelerator. John hasn't mastered this quite so well, and I have equal difficulty restarting it when warm.

Chainsaws are dangerous and are one of the reasons why the forestry industry has a high accident rate. It was sensible therefore to follow safe practice and only ever work with one if someone else was in the wood. For the coppice thinning this was mostly Douglas Clarke who brought along his own large saw. We worked hard, saws continuously whining and roaring as stems were felled, cut to length, trimmed of branches, and lop and top cut up. The deafening noise even blocked out the trains rushing past 50 yards away. By the end of the second day only about a quarter had been thinned; seriously adrift from the two-thirds rule. I should have been less optimistic, having been lopping and trimming trees since boy scout days, but it was the sheer quantity of branches and tops which coppice yields, especially the overmature hazel, that got the better of us. Every so often the cutting stopped, saws quietened down to tick over, and the accumulating debris was tidied and piled up. Thick straight pieces of sycamore and hazel, good for firewood, were stacked to one side for John's grate and that of Averal, my mother-in-law, though not for our centrally heated home. We did not finish thinning the coppice and natural regeneration before the great planting week was upon us; indeed it took two more winters of occasional days cutting and stacking, mostly with the help of our older boys, Jonathan and Stephen. Each winter since then we have cut, carried, heaved, and stacked 1 or 2 tons of firewood ready for use the winter after. Firewood needs to be thoroughly dry to burn well and give out much heat. Hence the maxim to cut each winter the next winter's wood. And, into the bargain, the wood warms you twice: from the exertion when it's cut and from its incineration when in the grate.

The big planting was finished by mid-April and just in time before the beautifully warm weather later in the month. The warmth brought on the trees and also the weeds. By mid-May the clean site was a patchwork of Dog's mercury, nettles, newly sprouting coppice, clumps of grass, and straggles of bramble. The scourge of *Clematis* had yet to reveal its entanglements. This assault of lush weed growth, freed from the fetters of overhead shade, threatened to choke the young trees struggling to get a hold. Sometimes this literally happens when *Clematis* entwines or herbage overtops, but more often it is their aggressive competition for moisture and nutrients from the same soil where the tender young roots of the new tree are trying to take hold. Trees are designed to last for years and years, but weeds have to make the most of every opportunity for their one season of glory. They outgun small tree seedlings in every department.

Even in plastic treeshelters the young trees could not outstrip the weeds. By late May it was time to apply herbicide in a 3-foot-wide ring around each tree. The chemical used was glyphosate, which is the active constituent of the garden weedkillers Tumbleweed® and Roundup®, and will control most types of weed, though after application it does take two or three weeks to show any effect. A single application in May or early June will give adequate control for the rest of the year. The same can be achieved by mulching or even hoeing out weeds, but both are too costly when there are hundreds of trees to treat. Mulching is a particularly useful way to control weeds for trees planted for amenity or in one's garden. Any inert material will do, such as bark or black plastic, though squares of old carpet are as good as any proprietary product.

Glyphosate is applied as a 2 per cent solution, and even though only a 3-foot-diameter spot was treated around each tree, about 35 gallons of water were needed for dilution. The knapsack sprayer had been bought with the gift from colleagues at the Alice Holt research station when I left to work for the International Institute for Environment and Development for two years. It holds three and a bit gallons. There was going to be much slogging back and forth over the site. But where could the large quantity of water be obtained? The wood has no stream or artificial water supply, the grey tank full of rain water was ruled out, being also full of slime that would clog the nozzle and denature glyphosate, and obviously, one couldn't commute the 15 miles to and from home. What was needed was a large receptacle in which to transport water. Margaret had the inspiration. Her mother's grey dustbin was languishing unused since Winchester had just switched over to wheelie bins. And, being the best of mother-in-laws she was, as always, more than willing to help.

A full dustbin and an already filled knapsack together held just enough water. The dustbin fitted snugly into the back of the Talbot Horizon and the garden hose used to fill it, but only after the dustbin had first been lifted into the car! When full it weighed close to 2 cwt or some 100 kg. Driving to the wood with a dustbin full of water at one's shoulder was a new experience. Every bump, every bulge, and every corrugation slurped water: never was such gentleness applied to accelerator and brake, never was such respect paid to corners and cambers. I should have remembered sooner the African solution, long employed by women bearing aloft their water jars, of a floating wreath of foliage or similar wave calming device. The dustbin lid alone was inadequate, even with Stephen or Ben holding it down, and a piece of plywood, nearly the diameter of the dustbin, floating on the surface cured the slopping.

In the knapsack sprayer the mixture was made up by adding 200 ml of glyphosate when it was about a third full of water. Pouring in the remaining water thoroughly mixed the herbicide. Margaret provided an old kitchen measuring jug for this bit of chemistry and by the time the sprayer was full the mixture boasted quite a frothy head. It positively foams if a proprietary mixture is added to make it more wettable, a surfactant which closely resembles washing up liquid.

The sprayer is worn like a knapsack as the name suggests and when full to the brim is heavy, awkward, and uncomfortable. It chafes the shoulders and rubs the small of the back, but at least it does steadily get lighter as the left arm pumps and the right hand directs the steady fine discharge. One filling would suffice six or

seven rows of trees and I perched my headgear, a U.S. Forest Service baseball cap, on the stake of the last tree treated with its peak pointing the direction I was going. The herbicide has no dye and soon dries without a trace, so it's easy to lose sight of where you've covered.

It took a whole day to treat the 1500 trees and was hot sweaty work in the warmth of late May. The protective clothing, which prevents skin contact with herbicide, prevented perspiration evaporating and fulfilling its job of cooling. By mid-June the weeds would wilt, turn yellow, and die. If application was done in hot weather the over-dressed operator would reach the wilting stage too! Weeds were controlled in this way for each of the first three years and then again in the fifth and seventh years after planting. The latter two weeding treatments were not necessary in the sense of getting the trees established, but undoubtedly they helped growth in just the same way that an orchard's apple trees each have their crinoline of bare brown soil to boost production. No more herbicide, or doses of any pesticides, will be applied in Taid's wood now for at least 100 years.[1]

The treeshelters required some attention and were not quite the problem-free invention that we had hoped for. From time to time a few needed straightening and their stakes hammered in deeper, though the shoulder of a 7 pound axe I would sometimes use is not the most sensible of tools. Weeds grew up inside some of the plastic tubes and smothered the enclosed tree, with grass and stinging nettles in particular offending in this way. To get the weeds out, the tube was first loosened from its stake and then raised about an inch off the ground. Then, using one's hand or just the index finger as a hook, in an action reminiscent of a lady freeing very long hair trapped when putting on a jacket, the tresses of grass and other tall weeds were steadily withdrawn from the treeshelter. The extracted weeds were not cut off but laid flat to await the shampoo of herbicide to kill them and prevent the problem recurring. Another concern while the trees were still small was scuffing or abrading of the soft leading shoots as they emerged from the top of the treeshelter despite the trumpet flare of the plastic. I tried sticking strips of foam rubber insulation to prevent the damage, but it didn't work. Fortunately, the trees soon stiffened and grew through this tender stage.

[1]Use of pesticides of any type in Britain's woods and forests is tiny, well below one per cent of the quantity that goes on to fields and gardens.

The growth of the young trees was a delight to a forester's eye. The first cherry had emerged by 20 June and by the year's end over half of the trees had grown more than 3 feet in height. Whenever at the wood in that first summer after planting, I walked between the rows and counted the newly emerged shoots, flagging aloft their achievement of outgrowing their plastic kindergarten. The ash and cherry grew best and especially well if they had been planted close to one of the old fire sites. The heat may have sterilised the soil and the ashes would have added nutrients. No fertiliser has been applied in Taid's wood. Indeed, it is rarely necessary to do so for trees growing on lowland soils, their modest requirements readily satisfied by the typical loams and clays. Preventing weeds from competing with trees is far more important for good growth.

Each year, in the autumn, a sample of the best trees are measured and their progress closely matches the best one can expect from ash, cherry, and oak on our ground. The fastest ash and cherry have averaged close to 3 feet per year for the first eight years. The thickening of stem diameter has been equally sterling, and in the fifth year we began what has become a time-consuming task, slitting treeshelters that were throttling the trees. The wild cherries first needed this attention. Their treeshelters were becoming club-footed like an elephant's, as stretch marks in the plastic at the base showed it was not degrading, even after 8 years, but was being forced out as each year's growth swelled the tree's stem. A Stanley knife proved ideal for slitting the tubes, by carefully inserting the blade to run up tangentially to avoid harming the bark of the very tree in need of release. Sometimes water poured from the slit where a well sealed treeshelter had filled with rain, and occasionally a vole had found the mini-greenhouse an ideal home. Usually stones and soil squeezed the root collar adding to the danger of girdling. When this happened it was important to work the shelter loose as well as slitting it.

The slit tubes were left still attached at the top and attached to their stake for a couple more years to deter fraying by roe deer and nibbling of bark by voles and mice. Sooner or later the treeshelter must finally be removed and the litter of plastic gathered up. The failure of the plastic to disintegrate, or biodegrade, is probably due to an excess of ultraviolet inhibitor added during manufacture. All plastics exposed to sunlight eventually break down, but this process can be long delayed by incorporating plenty of carbon and specific chemicals to prevent the ultraviolet component of sunlight doing its stuff. It's rather like factor-25 barrier cream. In the countryside, whether it's old fertiliser bags or outgrown treeshelters, one wants waste plastic to degrade. The one means of disposal which doesn't help is to bury the plastic, hiding it not only from view but the all-important sunlight. Care is also needed not to leave plastic scattered or blowing about as an offence both to the eye and to the stomach of livestock, particularly cattle, which may try to eat it.

From about the fifth year the most vigorous trees, by now young saplings, were ready for formative pruning. Forks, misshapen growth, and large side branches are cut out to encourage a straight stem. With the trees planted about 10 feet apart, they were not close enough to shelter each other and suppress growth of larger side branches and, unlike most conifers and poplars, only a low percentage of the oaks and ash were naturally straight. This job for secateurs and long handled pruners is rarely mentioned in textbooks; writers usually hasten from weed control to the cleaning stage when unwanted growth is cleared out. Yet it is clear that the finest stands of oak and beech we admire today owe their impressive quality to formative pruning 100 years before. It is easy Saturday work and immensely satisfying to prune and conform nature to the 20 feet of straight trunk beloved of sawmillers.

The early removal of large side branches keeps knots small, but should all branches be removed up to a particular height at this early stage? William Pontey, writing on tree pruning in 1808, definitely thought not. He was the Duke of Bedford's nurseryman and tree planter and felt strongly enough to write a book called *The Forest Pruner*. For him good pruning was to thin out all large limbs, that is the ones three or four years old or more, and keep the small ones to provide leaf and so help growth. The Forestry Commission has recently begun to research this systematically and investigate other ways of formative pruning since the operation will be increasingly important, owing to the upsurge in broadleaved planting of trees spaced rather far apart. In the wood most of the oak required formative pruning along with some ash. Oak rarely grows straight of its own accord, especially in the open, and pruning to train it upwards is invaluable. Ash has a strong tendency to straight growth, but all too often it is thwarted by frost damage to freshly opened shoots or the burrowing of the ash bud moth into the topmost bud, causing it to die. Both cause forked or multiple stems, and Taid's wood has its share of ash looking like giant pitchforks. There's also another cause of forking in ash. In June the expanding, but green and pliable leading shoot, is easily snapped when chosen as a perch by an overweight wood-pigeon. Often ten or a dozen shoots are so snapped by mid-summer.

Wild cherry requires little formative pruning. It grows vigorously with a tall straight stem which rarely suffers dieback or breakage. But cherry does develop whorls of heavy side branches which need removing before becoming too thick. Pruning is the only way if the trees have been planted far apart, and it must begin when trees are as young as five or six years. This was so in our case and each year many cherries have been pruned. It is mid-summer work since the tree, like ornamental garden cherry trees, is susceptible to canker. By pruning in the months of June and July infection is usually avoided. Even so several cherry trees have developed canker and in February, when it's most obvious, the gummy, brown, gelatinous mass exudes from infected pruning scars, branch crotches, or wounds looking like blobs of jelly. When thinning begins these infected trees will be the first for the chop.

The cherries had also suffered in most years from a defoliator which stripped trees of freshly opened leaves in the spring. The cause remains unclear. The leaves are nibbled or chewed nearly to their base leaving a crescent of green resembling a 5-day-old new moon. Black cherry aphid infestation has curled young leaves at the branch tips in most summers and appears more unsightly than harmful. In spite of these pests the cherries have grown vigorously.

At the same time as pruning, *Clematis* is cut out or unwound. All over Taid's wood rooted snakes of this climber have become established and they sway in the breeze, like the rattans of the tropics, to clasp unwary trees and build themselves on another's robustness. This will be a long battle on our calcareous soil. The aim is to defeat the climber, though I have no enchanter's skill to charm it away. Indeed, *Clematis* has an uncanny, almost magical, knack of disguising exactly where it's rooting, and hence where it should be cut, while displaying for all to see its entanglements of streamers festooning the shoots and branches of the trees we are trying to grow. In only a few years the woody stem of *Clematis* becomes camouflaged in moss, but more irritating still is its habit of rooting in a clump of nettles or underneath a low bush. This isn't by chance, nor is it some dastardly plan of *Clematis'* own making (it hasn't scaled the heights of intelligence despite getting the better of me many times). No, this knack of hiding arises from young shoots being protected from browsing and doubtless sheltered from herbicide spray as well. They are overlooked, waiting to arise when danger is past.

As I write, the young trees are in their eighth growing season and one maintenance job remains before the stand is left until its first thinning. This is the work of cleaning which, as the name suggests, is the removal of all unwanted woody growth so that the chosen trees grow unhindered. In Taid's wood the near impenetrable throng of shoots of young sycamore coppice will be thinned out, as will some but not all of the sallows, birch, and hazel that have seeded in. Smaller weeds are now less of a problem and colonisation of wild flowers is encouraged including primroses and, if possible, cowslips in the spring, and the knapweeds beloved of butterflies in the summer. These fine purple-flowered plants include 40 closely related species. They attract many spectacular insects and my father was thus attracted to them too, but taxonomically the plants always remained for him (and me) simply 'difficult hard heads'.

The most trying of cleaning tasks in Taid's wood, and the job which perhaps the Forestry Commission had also found daunting when establishing the pine and beech, will be coping with the *Clematis*. Hopefully, attention to weeding and tackling it while pruning will have kept it at bay. Already all 1.3 hectares have this tiresome weed and I shall simply have to persevere to keep the trees unfettered and free. That is the forester's commitment to his crop, to his profession's offspring, as he applies to trees God's lesson for Adam that 'by the sweat of thy brow shall the ground bring forth food'. Like roe deer, which browse just about anything except *Clematis*, I find even brambles infinitely preferable to this climber.

8

The half-past-two roe deer

During visits to check the new planting in the summer of 1987, roe deer were frequently sighted. Only after several visits did it register that each time it was the same graceful creature that was disturbed from its early afternoon browsing, and in almost exactly the same place. For several weekends and occasionally in mid-week, the deer and I would keep this 2.30 appointment, though I was the one who usually failed to turn up. Perhaps the quiet of the wood and lack of human intrusion helps explain its clock-like regularity, but roe are renowned for good time-keeping as they work their territory.

Roe deer are woodland inhabitants which rarely venture far from cover, preferring young plantations and thicket to the darker windier conditions of older stands. The depredations of wartime fellings, the expansion of young forest in the uplands, and the patchwork of new woods and resuscitated copses in the lowlands have, today, proved ideal habitat for roe. They are abundant now in most counties and are a threat to much new planting. Roe deer, one of our only two native deer species, are found throughout much of Europe, including eastern Europe and the new nations once part of western Russia, but conditions have not always been so favourable. The species had all but died out in the English lowlands by the 1850s and the present exploding numbers owe ancestry to Scottish stock, and escapes from parks such as Petworth in Sussex. Today they occur widely in southern and western England, Norfolk, and northwards from Yorkshire and Cumbria including the whole of Scotland. They are still scarce or absent from Kent, Wales, and much of the Midlands.

On most visits to the wood roe deer are seen, and often in pairs. They probably hear or scent human presence long before one catches a glimpse, and then only the whites of their hindquarters as likely as not, being wary animals ever on the alert. In winter, when less cover reveals them more easily, their coat is a dullish grey or brown. This colouring develops in September or early October as long winter hair grows through their short summer fur to camouflage the sleek foxy red. This transition, like our own adjustment of clocks back an hour at the end of October, occurs almost overnight to herald winter. Return to the summer dress of glistening reddish brown is altogether slower, perhaps again sharing our uncertainty when warm spring days so easily tempt, deceive, and ultimately mislead over the imminence of summer. Tufts of their old winter hair fall out intermittently over a 6- or 7-week period in May and early June.

Adult roe are quite easily seen at most times and, as I learned, are remarkably regular in habit, only changing routine in hard winters or hot dry spells in summer. They are most elegant animals, poised and only rarely breaking into flight. They are equally refined in diet: dainty feeders nibbling a shoot here, browsing a leaf there, and all the time garnering moisture from the foliage itself and especially dew at breakfast time. They do have some preferences, brambles over the toxic *Clematis* being one of them. And, although they do graze, their strong preference for herbs gradually alters the balance in favour of grasses along rides and on the woodland floor in woods frequented by roe.

This variety of diet threatens the forester's newly planted trees, offering, as they do, a new taste to the fussy roe exploring the woodland delicatessen. There is quite a lot of evidence to suggest that young trees just after planting are more at risk than ones which have grown up naturally, even if of the same species. It is possible that the stress of planting, and during the ensuing weeks while the tree becomes properly established, causes physiological changes which browsing animals can detect. It's as if the young seedling, fresh from its nursery, unwittingly advertises its arrival on site and the deliciousness of its well nurtured, healthy foliage as the latest delicacy. And roe are drawn to tear or pluck untidily the proffered shoots. Like all deer they have no upper teeth, only horny pads for gripping rather than biting. Sheep, hares, and rabbits, with better furnished jaws, clip and browse cleanly. This distinction helps identify the culprit when lamenting over a freshly decapitated young tree.

A deer's lower teeth come into their own for bark stripping. Fortunately, roe are the least of offenders among Britain's deer of what can be serious damage, especially by red deer in upland forests, although they have no such diffidence about fraying. This is a spring-time activity when stems, called fraying stock, are used by the deer to clean their antlers of velvet. Later on male bucks will do more rubbing of stems to deposit scent from glands on the forehead to mark out their territory in readiness for the mating season or rut. Fraying stock is usually woody bushes or small trees (sallow is a favourite) and all the bark is rubbed off or left messily shredded. For fraying roe tend to select unusual or different trees from the main type present. Their ability to seek these out is uncanny. After planting the 1400 or so trees in Taid's wood six specially selected wild cherries were added. They came from the Forestry Commission's research nursery, and I paid £2 each for them. They were surplus trees from an experiment which was testing selections from a French tree breeding project. Three of the special cherries were planted at intervals around the edge of the new planting and three in a group to fill a gap elsewhere in the wood. Since the trees were already nearly 6 feet tall, and well above a roe's maximum browse height of 3 feet, I dispensed with use of treeshelters. They were planted on a Saturday, and by my next visit, late the following week, all except one had been frayed! The roe deer relished the French stock. Two of the trees never recovered despite cosseting by belated enclosure in a treeshelter. The other kind of tree in the wood which they like to fray are small Douglas firs becoming suppressed by bigger neighbours, and they always go for the one yew near the entrance which has struggled against other odds to make 10 feet of height under the dense shade of unthinned beech. At least they have never resorted to bark stripping.

We do not cull the deer which pass through the wood. The lease allowed us to but we have decided to protect the young trees individually instead. Deer pressure on ground vegetation is evident but not excessive, even though in one year one of the only two patches of cowslips clearly became one tasty mouthful. Some culling may have been done, since a deer seat was erected just outside our boundary with the sighting line over the newly planted trees of Taid's wood, which is a warm sheltered spot where deer would browse. Our climate is not always congenial and although roe can tolerate cold harsh winters, cool wet weather in April and May when does are giving birth can, sadly, be all too lethal for both doe and fawn. Records show that such weather conditions lead to high mortality, in effect a natural culling of the population. Only on a couple of occasions during the ten years of ownership have the carcasses been found in the wood. Certainly the most recent, in late spring of 1994, coincided with the very wet winter and spring of that year. Other roe quickly replace the loss as defence of territory is resigned by death. For owners and visitors alike, pausing, breath-held, a passing buck or, better still, a doe with fawn will continue to thrill as these most graceful of God's creatures are glimpsed, even if not always at half-past-two.

Beyond their attentions to trees, roe deer bring a second and very different problem to the wood, and one to which the trees are quite immune. They carry ticks which brush off onto foliage

as they move through the undergrowth. The ticks wait to hitch a lift on another passing beast to plug in, limpet-like, to a new food source. Humans are perfectly acceptable beasts! Ticks actually sense a human's presence from vibrations or heat; they can't see at all, being completely without eyes. Ticks are tiny creatures of the spider family and appear like extra-dark freckles on one's skin. Watched carefully these freckles move, but it does help to know the geography of one's moles. In late spring and summer ten or a dozen may cling to my trouser legs at the end of a morning's work. Each is carefully flicked off, though one can never be quite sure all have been disposed of. More intimate body inspection is called for at bathtime since these tiny mobile limpets painstakingly explore every recess and fold of the skin to discover the softest, most tender parts. There they insert their mouthparts and inject an anti-coagulant to ensure a steady flow of nourishing blood. In time they begin to swell, and a grey sack distends like some unwieldy rucksack. Unlike roe deer and their taste for variety, the tick knows the one meal it wants and gets stuck in.

Swollen tick greatly magnified. ● Actual size when 'caught' in the wood: tinier even than 'Small' in Winnie-the-Pooh.

Even by the gorged stage little pain is felt, only slight irritation perhaps. Normally one finds the offending tick long before it's so obvious and it is extracted. It's best to do this with a pair of tweezers held as close as possible to the skin and twisting the tick anticlockwise without pulling or jerking. Experts dismiss cigarettes, lit matches, alcohol, and other amateur remedies, so tweezers should be in the first-aid kit. After a few twists the tick comes out, but it is important to make sure this includes its embedded teeth. If not there is a slight risk of catching tick fever, caused by the louping ill virus, which can kill young grouse and lambs and occasionally upset humans. Recently the much more serious Lyme disease has arisen. It is named after the Connecticut village of Old Lyme where distressing cases of child arthritis attracted attention in 1975, though it was not until 1981 that the tick was identified as the carrier for the spiral bacterium responsible. In Britain Lyme disease is rare, but it has turned up in and around the New Forest and in East Anglia. Fortunately, very few people have contracted it. Nevertheless, for summer visitors to the wood, tick inspection is recommended on departure and before going to bed. So far no one, not even myself, has contracted tick fever and certainly not Lyme disease. Occasionally, though, ticks have escaped initial detection and once or twice have only been discovered as they began to swell.

There are 18 kinds of hard ticks in Britain and it is the 'sheep tick' which carries Lyme disease. Sheep ticks will quite happily attach to humans, and dogs and cats in particular. Both people and pets can contract the disease, but it is entirely curable with antibiotics once it is diagnosed.

The wood, of course, boasts more wildlife than just deer and ticks though these two are the extremes in animal size one usually encounters. In between come woodmice and moles which inhabit the litter of the woodland floor or burrow just beneath. A delicately grey furry mole surfaced one warm afternoon and hurried along beside my parked car, paying no notice as I watched its progress. This is the only time I have seen a mole in the wood though evidence of their underground activity is commonplace. Most noticeable is the sudden appearance of a narrow ridge across the track or ride where the mole has tunnelled from one side to the other. One near the firewood pile in Taid's wood has been constructed each year in the spring at almost exactly the same point. This molish excavation has gone on for several years despite the regular flattening which the passage of the car and British Rail's assortment of vans must impose. In fact moles inspect their entire network of tunnels every few hours, repairing any damage as they do so.

Moles tunnel to find worms and caterpillars and other meaty dishes. The tunnels also act as a trap for worms to keep a mole's larder filled. The mole senses the worm by the slightest of vibrations or sounds and nips it, paralysing the worm to stop it crawling away. Moles hunt for about four hours each day to find sufficient to eat nearly half their body weight every 24 hours. But moles also have a vegetarian bent. They have been known to tunnel along a line of newly planted trees and eat away the roots, killing all the trees over a 30-yard length. Fortunately their taste for tree roots is never sufficient to make them a serious forest pest.

Woodmice and voles, like moles, are much more to be welcomed than despised even though they will gnaw the tender bark of young trees. They have done this on the trees in Taid's wood, particularly the ash, when the treeshelters have been removed or slit open. The damage has not been serious and I suspect rabbits have joined in too. Voles are common in open woodlands and, in our case, doubtless provide good fare for the tawny owl which is heard most nights whenever we camp at the wood. They could have been the attraction for the pair of buzzards Mike has seen recently, gliding lazily and crying their haunting, almost plaintive, *'peeeiouu'* call, and for the one sparrowhawk seen in the wood, though, as its name suggests, this raptor (bird of prey) mainly hunts for other birds—passerines such as songbirds and finches. It may even have helped itself from the flock of long-tailed tits seen from time to time on the railway fence, or a tiny goldcrest as they emit their excited squeaks in A-sharp, hurrying from pine to fir and back again in search of insects and spiders. The much larger female sparrowhawk can cope with altogether more substantial fare and will prey on birds even the size of wood-pigeons.

Wood-pigeons make their presence felt on most visits. Often when walking through the wood a sudden flurry of flight startles as the noisy birds, rustling leaves and clapping wings, take to the air. At dusk on a summer's evening the noise can be quite frightening when a flock returns from a day's feeding on whatever crop is in the adjacent fields, and moves around the woodland canopy looking for sleeping quarters. When our boys were young, such strange night-time noises kept them alert and on edge as Dad tried to settle them for the night under canvas. Indeed, they almost matched the nearby trains for volume, magnified, as it always seems, in the hours of darkness. Wood-pigeons take a lot of grain and seed from farmers' fields, though their favourite diet of cereal grains makes them thirsty and they are often seen drinking from puddles, ponds, and cattle troughs. Their numbers have grown in recent years; at least this is the impression one has in the wood. Perhaps the beech trees are becoming a more congenial home. Certainly Gilbert White records in his notes about Selborne's natural history that wood-pigeons have an affinity for beechwoods. They remain little loved in the country-side, as they do their bit to keep down grain mountains(!), despite the caressing and soothing coo of their song. I think, too, they raid the keeper's grain put out for the pheasants, but so do other less endearing inhabitants of the wood.

I regret to say that rats have taken up residence. They are attracted by large quantities of wheat dispensed throughout early winter from 44-gallon drums to feed pheasants. At least one of their homes has been under the grey water tank only a few yards from where the nearest feeding drum stands. Happily they are not often seen, but neither are they easily eradicated. Sadly the one fox lair we had is no longer occupied, so the rats feel safe.

We have no badger set within the confines of the wood. And, without any pond, there are no wildfowl and few bats, and amphibians are only rarely seen. Nor has the Basingstoke lioness, sighted 7 or 8 miles to the north of us in September 1994, made an appearance, real or imagined, like it did in the pages of the local press.

In safer if less interesting contrast, both rabbits and hares inhabit the wood throughout. The hares are much the more visible with their large black-tipped ears and loping gait. Their hind legs move in a curious unison causing the hare to pitch back and forth as it lollops slowly along one of the rides. Their huge bulging eyes appear alert to everything, but are at their best at dawn and dusk. When suddenly surprised the hare smartly weaves its way between the trees as one would expect. But, in spring and summer the sea of Dog's mercury on the woodland floor both hides all but the hare's ears and then fails utterly to provide cover as loud rustling betrays any movement by the frightened animal and the waves of green slow its speed and track its progress like a ship's wake. Hares are thought of as creatures of the field running eccentrically to escape danger, and not only in March, but in recent decades they have become more and more woodland inhabitants. Certainly they appear at home in the wood and are a delight to see.

One is somewhat less generous towards the hare's distant relation, the rabbit. In an earlier chapter I described the damage they cause and the huge number of burrows, over 50, found in the wood when we bought it. The law requires us to control rabbits to avoid our land harbouring a nuisance for our neighbouring farmers and growers. Mike and Annie with their organic vegetables would heartily agree. Indeed, even our best efforts at control have not eliminated the rabbits, and Mike has erected a rabbit-proof fence down the length of our common boundary. I hope it has kept the rabbits out of their land, but it appears to have kept in our wood a pullet which turned up in the winter of 1993. For several weeks the young chicken strutted anxiously up and down the fence as if on guard duty. It then disappeared, but a couple of months later resumed its patrolling and was later joined by a cock. Neither Mike nor I know where they have come from.

Except in the first year when Alistair helped, John and I have carried out rabbit control most Novembers. The days are short, grey, usually damp, and not for lingering: a long half-day would see the job through. The work can be done at any time in the winter though tradition used to say that rabbits were hunted when there's an 'r' in the month. We begin by searching systematically for burrows, especially along the boundary. Once one is found others are looked for in the vicinity. A single entrance is rare, usually there will be several even for a small warren. The law very carefully stipulates what to do next and how control should then be carried out.

I don't like rabbiting even though the method has been precisely laid down and is as humane as possible. But rabbits are not native to Britain, they are increasing in number to their pre-myxomatosis levels, and they do cause damage to young trees and a great deal of harm to farm crops. Mike certainly wants rabbit numbers kept down—one can never eliminate them—as no totally rabbit-proof fence exists. Even 3-foot-high fences with their bottom netting carefully folded out to a distance of 8 or 9 inches to prevent burrowing underneath can be breached. Foresters have seen rabbits, when pressed, climb over 6-foot-high wire netting. In snowy weather their tracks have been seen arriving at a fence and continuing on the far side, and there's not a hole in sight, only freshly browsed tree seedlings and other signs of mischief. Such mischief is not new. Two hundred years ago Gilpin, in his *Forest Scenery*, complained about the damage they caused, they '. . . nipped in the bud the glory of England preventing the growth of oak trees which were so badly needed by the navy for shipbuilding.'

But for all these complaints, rabbits are a delight and have been a part of our countryside nigh on a millennium. And, they do clip the grass to a fine turf while avoiding the very flowers we enjoy, rarely if ever browsing violets, primroses, poppies, and periwinkles. Nevertheless, the law places a duty of control on responsible landowners. Both the Pest Act of 1954 and the Forestry Act of 1967 labour this point. The rabbit can do much harm, and so some control must be exercised. However, even their worst is no more than a local problem and doesn't compare with the destruction wrought by a storm.

9

No pen could describe it, nor tongue express it, nor thought conceive it unless by one in the extremity of it.

Daniel Defoe on the Great Tempest of 26 November 1703

The great storms of '87 and '90

A motel bedroom in Canberra, Australia, was a safe place to be when the storm of 16 October 1987 struck the home counties of England late Thursday night and early Friday morning, but not good for one's anxiety. News of the ferocity of this hurricane was relayed with gathering detail and urgency to the antipodes throughout their Friday evening and Saturday. I may have been one of the first in Australia to hear about it since Mike Duckering of Tear Fund phoned the motel at a pre-arranged time about a proposed visit to Ethiopia in November to help assess the resurging threat of famine in that sorely pressed land.

Mike phoned at 7.10 p.m. Australia time, 10.10 a.m. Friday morning in England, and just four hours after the height of the storm. He was using an emergency phone, the only one still working in the whole of Tear Fund's offices, and he sketched in the scene of devastation and destruction across the south-east.

I phoned Margaret straightaway to hear, thankfully, that at home only the fence between ourselves and the Allens was down and that already David Allen had kindly offered to put it back up. Our two younger boys, Stephen and Benjamin, couldn't get to school in the village of Bentworth because of fallen trees blocking the roads and no electricity. I tried to get through to my mother in Petts Wood, in north-west Kent and right in the path of the storm, but there was no reply. It wasn't like her to be out at that time. Her phone rang and rang. Hours later I got through to my sister who reported mother well, but one of the two great oaks in the garden—survivors from conversion of Towncourt wood to housing development in the 1930s—had had its top blown off. My mother's phone remained out of action for days.

This anxious personal assay of the damage was augmented by Australian news and TV broadcasts. Helicopter shots of flattened woods around Hastings, trees uprooted at Kew, houses without roofs and caravans blown hundreds of yards, made graphic footage. I devoured news programmes to glean the extent of the damage. What had happened to the wood, bought only two and a half years before, and evidently located on the western edge of the worst of the storm? I felt embarrassed to ask, people being far more important than things, but every British forester knows too well the havoc gales and storms cause. Britain is the fifth windiest country in the world, and wind blowing down or snapping great trees as if they were matchsticks is the most serious destruction that trims the yield of our vigorous upland plantations. Many such plantations grow on soggy peaty soils on exposed hillsides and are therefore especially prone to wind damage, but prolonged and very heavy rain had preceded as well as coincided with the great storm of '87. The soils were thoroughly wet. Was the wood now like crops I had seen: uprooted, a tangle of trunks and branches, a mess reminiscent of the desolation of the Somme or Verdun?

Margaret had told me that the storm force was reportedly off the scale and the ground as soggy as feared. And I knew, musing in the motel bedroom, that the 60-foot-high pine and beech were ripe for throwing down. A colleague at Australia's forestry research institute discouragingly remarked that the destructive force of wind is in proportion to the cube of its speed: any gust over 70 m.p.h. will damage or uproot trees. These gloomy thoughts were uppermost next morning when my spirits should have been quite otherwise. Ken and Marian Eldridge and Chris Harwood, international experts on eucalypt trees, were taking me up the Brindabella mountains behind Canberra to see the very snow gum trees from which seed had been collected for trial plantings in Britain a few years before. These trees were exceptionally tolerant of cold and their offspring in England had survived the exceptionally low temperatures in the winter of 1981/82. Against the odds, they had coped with the extreme cold, unlike British Rail's engineering marvel, the hapless prototype advanced passenger train, which had not. The day on the Brindabellas was crisp, the light brilliant, and the gums impressive at their high mountain tree-line with snow still ankle-deep covering their roots. The pristine forest, clarity of air, and white brightness all around failed to dispel the brooding. Even the Lamingtons, sponge cakes generously encapsulated in chocolate and coconut, didn't help much as we picnicked. The evening news from England brought no relief and I continued to scavenge papers for every detail of the storm to learn whether central Hampshire had been hit as badly as my native Kent.

On Monday the world's stock markets crashed. Tuesday's headlines of that Black Monday are now history, but the coincidence of great storms and great crashes reminds of our frailty, whether supposedly under our control or not. On the Tuesday I moved from the motel to the Harwood's home. Chris and his wife Lucy from Tonga made me welcome, encouraging me to phone home to England straightaway. By this time I had pretty well convinced myself that the wood couldn't have gone completely otherwise Alistair would have phoned Margaret and she would have tried to get in touch. Anyway I got through and told Margaret the new contact number and enquired light-heartedly how she had weathered the stock market crash (we hadn't got any shares), only to hear her report that that was the least of her worries. What did she mean?

The heart raced, a sinking feeling enveloped, and resignation to the inevitable all began in the split-second delay possessed of long distance calls that go 22 000 miles out into space and back. Was she preparing me for the worst? Had I deluded myself, unwilling to accept what had obviously happened? Reason insisted it was the hurricane of the century. The pictures on television clearly conveyed this message. Why should we have escaped? Satellite communication's momentary lag expired. Jonathan had badly twisted his ankle, jeopardising the next week's walking with grandpa—the woodland was not destroyed. My dear wife has her priorities right, as people matter ever so much more than things. She kindly followed up the news about Jonathan to say that John had gone to the wood on Saturday and found a few big pine trees down and one or two of the newly planted trees in their plastic shelters somewhat askew. There was no devastation. This investment, unlike that of shareholders, had largely escaped.

I needn't say that this was a relief, as the above baring of my feelings reveals. I was thankful for John's evident interest and, too, toward God for His providential dealings. The last remark sounds selfish. After all what about all those who didn't fare so well, or what about emaciated boys and girls, or the very old starving and uncared for in Ethiopia? I cannot give an adequate or complete answer, but such questions, so commonly asked at times of calamity, shouldn't deny us giving thanks for all the good things we do receive. It seems to me we shouldn't stop giving thanks because at other times, in other circumstances it might have been very hard to do so. And, of course, it shouldn't stop us from always helping others whenever we can.

On returning from Australia at the end of October I became aware of having missed the drama as well as the hurt of the great storm. I had also missed something else. The morning after, when all around was destruction and havoc, brought out a camaraderie, a Dunkirk spirit, between people who before were simply acquaintances or neighbours limited to exchanging pleasantries. For once the weather claimed the whole conversation and not just the preliminaries. People helped one another in their shared and very real needs. As others told me about this collapse of English reserve, it recalled a similar day when commuting by train to school in London. On the day after Eric Lubbock won the Orpington parliamentary seat for the Liberals in 1962, the carriage was noisy with chatter and talk about this unexpected twist in politics. Animated conversation passed between complete strangers normally content to hide behind the morning's *Telegraph*.

I had missed one more thing as well: the shock of waking on the fateful Friday and finding one's surroundings shattered—a familiar tree uprooted, a much loved woodland walk blocked and impassable, a local landmark gone. For many in the four home counties of Essex, Kent, Surrey, and Sussex, this shocking loss made a deep and lasting impression: it was heartbreaking. My mother felt this bereavement and wrote the following lines a few days after the great storm in order, as she put it herself, 'to ease her mind'.

Kent's trees
Autumn 1987

This year few leaves will turn
Crimson or gold. Wind-burn
Shrivelled them ashen grey
In one wild night. The day
Broke upon trees flung down,
Uprooted, splintered, crown
Dragged to the earth, limbs torn
From trunk—age worn,
Age-hallowed, in their prime:
Gone, gone, so many gone, before their time.

Joyce F. Evans
20 October 1987

The great storm was exceptional, but it was not unique. Less than three years later one almost as severe struck central southern England, as had the storm of 1703 which was, as Gilbert White tells us, talked about for years afterwards and even by those who, like himself, only knew of it from parents or the recollections of older friends. Gilbert White, like Daniel Defoe, described it as an 'amazing tempest' which 'overturned huge spreading oaks'. Not unique either is the anguish at the sudden destruction ferocious storms inflict. Jane Austen, looking out of the Rectory window at Steventon a mile and a half from our wood, and rather more than a century and a half ago, observed with distress the dreadful storm of Sunday, 8 November 1800 and wrote later saying it '. . . has done a great deal of mischief among our trees'. Several elms had been thrown down in the grounds yet she reflects thankfully that 'No greater evil than the loss of trees has been the consequence of the storm in this place or in our immediate neighbourhood. We grieve therefore in some comfort.' But grieve we all do at sudden loss.

The tally of damage in the wood amounted to eight pines, seven Douglas firs and five birches uprooted, and tops blown out of two oak and two beech. The only larch was also down, confirming the surprising susceptibility of this tree, as the National Trust had found in their survey of storm damage in the Willet memorial wood in Petts Wood. Many branches were broken or torn from the clasp of their tree. We came off lightly indeed.

We were not so fortunate with the late January gales of 1990. At the peak of the storm, early on the afternoon of the 25th, across the field from my office at the Forestry Commission's Alice Holt research station great oaks, which had been planted after the Napoleonic wars, were swaying and snapping like twigs. Right outside the window a small stone pillar buttressing a parapet cracked, feinted, and relinquished duty after 170 years at attention. A few minutes later staff were allowed to go home early because of the weather.

Setting off cocooned in the car and anaesthetised from the severe storm all around, I decided to go to the wood to see what the damage was. At Ellisfield the road was blocked by a car, ahead of which was a Volvo slowly creeping under a great uprooted fir just hovering high enough above the lane to allow it to pass. The delay was long enough to see a second fir crash to the ground in the wood

nearby, but not long enough for me to come to my senses and realise the foolhardiness of driving in such conditions. The wood was never reached that afternoon. All roads to it were blocked. I arrived home safely at about 4. 30 but was coy about my stupidity. On the phone that evening Sally, who can just see the wood from their upstairs bedroom, thought it seemed all right, at least from that distance.

The wind was still blowing tenaciously when a visit was eventually made on Sunday afternoon, 28 January. This wasn't really work, but more like the example Jesus gives of helping a donkey on the Sabbath which has fallen into a pit. Leaving the car at the gate I walked around the wood as best I could. It was an obstacle course of trunks and fallen branches. Both cross rides were blocked by fallen pines, now horizontal and hinged to newly exposed root plates bright from the white chalk subsoil suddenly levered into view. Some hovered leaning steeply, unwilling to concede the final indignity. Two Douglas firs were snapped at mid-stem with long red splinters left flaming upwards like some cruel fairy queen's crown. At one point on this tour of inspection, thoughts distracted and tripping over a log, my purse and I unwittingly parted company. It was gone when I got back to the car and has never turned up since. Also gone were 115 trees, mostly pines and mostly the best ones including the two biggest of all which had achieved a massive 41 cm (over 16 in.) in diameter in their 33 years of life. These champions had the biggest crowns of branches and so were exposing the greatest sail area to the wind. A number of good Douglas fir were also down. The voice of The Lord had truly 'twisted the oaks and stripped the forest bare' as Psalm 29 likens the power and authority of God's word to a wind of very great force.

The gales continued into early February and during a quick visit on 10th several more pines were found uprooted or leaning. One was propped diagonally across the main track just near the water tank: it didn't finally come down until the following autumn. The one surprise was that in Taid's wood only one or two of the young oaks, and none of the ash or the cherries despite their good growth, were even leaning, and no trees at all had been blown over.

The wood was insured and the fee at the time of £40 per year covered fire, storms, and devastating pest or disease outbreaks. A claim was prepared for the lost timber. The diameter of each blown pine and fir was measured near the base and a sample of nine were measured along the timber part of the stem to calculate their volume. These data were used to estimate the total quantity of timber blown. It added up to about 44 cubic metres of the best trees, or about 40 tons or rather more than 1500 cubic feet of timber. We sent the claim off to Frizzells, our insurers, with the estimate of the timber value. This can be done either by computing what the trees are worth, or are going to be by discounting their future sale value to the present, or by summing all the costs involved in planting, growing and caring for the trees so far. Our pines and firs were getting close to final felling, so we valued them as a crop using the first method.

The insurance assessor from McLaren, Dick, and Co., a firm of chartered loss adjusters in Kent, visited the wood on 16 March and agreed the valuation. It was then we discovered that we were not covered for the first £1000 of each and every loss. The £393 left of the claim was meagre recompense for the loss and cost of clearing up the damage.

Since 1990, autumn and winter gales have continued to throw down the occasional birch and decapitate beech trees sick from beech bark disease. This disease, common with beech on chalky soils, kills the bark at ten to fifteen feet up the trunk. A year or two later the dying or dead tree snaps at this point of weakness. It mainly affects beech in young middle-age when 20–40 years old. Control is problematic though one can scrub the bark of infested trees. The white woolly glazing of badly affected stems shows the ones to work on, but it is difficult to get high enough, even with a ladder, and the job is unpleasant as a gentle rain of fine waxy flakes mixed with dirt from the trunk falls on hair and anorak and inside one's collar. And, I am not convinced that scrubbing off the beech coccus, which causes the white waxy scale and stresses the tree allowing the fungus to attack, really does work.

The wind which spared us great devastation claims instead an annual surety. Each year we lose two or three beech, as tops snap off. The wind reminds us gently but persistently of its potential to destroy and lay waste. The point was well made and we took it to heart.

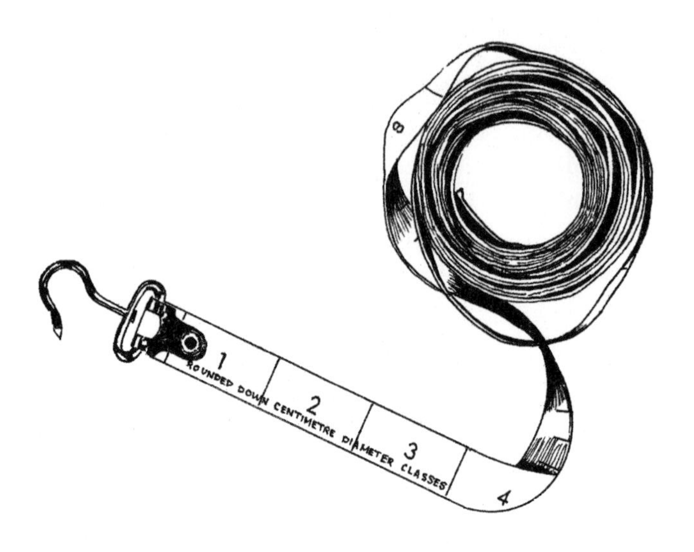

10

The timber sale

The two great storms revealed how vulnerable the pines were. Their tops rose above all but the tallest of the beech and we couldn't afford further attrition to this most valuable part of the wood. In the spring of 1990 we decided to sell them all.

Two other considerations helped make up our minds. Several of the pine trees were becoming chlorotic with pale needles and poor prospects for many more years of growth. Measurements of the distance between branch whorls at the tops of the windblown pines revealed that even on healthy trees each year's growth had slowed from over two feet per year in the early 1980s to only a foot or so in 1988 and 1989. The chalkiness of the soil was taking its toll on even the tolerant Corsican pine. Secondly, the three rows of beech between the double rows of pine urgently needed thinning yet individually the beech were small and, on its own, the job would be unprofitable. If all the pines were felled, the beech trees next to where they had stood would immediately be released and enjoy extra growing space. Rather like thinning out seedlings in a flower bed, the remaining ones are greatly helped; unlike the flowerbed one cannot pot on or plant elsewhere the thinned-out trees!

By selling the remaining pines from all 5.5 hectares on which they grew, a sizable and, we hoped, attractive parcel of timber could be offered to the market. Also, by doing it in one operation, it would be easier to fell the trees and to cut and extract the logs. And, there would be no need to worry about treating the cut stumps to prevent a serious fungal disease, called Fomes, infecting the remaining pines. When thinning pine, especially on chalky soils, this precaution is essential to avoid killing some of the trees left behind as the fungus turns parasitic. By cutting them all there would be none left to worry about. Fortunately this disease is of little consequence to beech or any broadleaved tree.

The main disadvantage was one of timing. Britain was entering recession, the 1990 gale had brought additional timber on to the market, and thus the price for timber was falling. It was difficult to guess when prices might eventually pick up, but we weren't obliged to sell if the best price offered was really not enough.

Timber is usually sold in one of three ways. If the seller knows the market well he can negotiate privately with a buyer, but he must have a good idea of the value of his trees. Placing the timber in an auction is a useful alternative which will realise the market price though it is perhaps less suitable for a one-off sale. The third way is to invite tenders for the timber by sending out particulars to firms who might be interested in buying. We opted for this approach and also chose the easiest course of selling the trees 'standing', that is the buyer does the felling, extracts the trees, and finds markets for the timber produced. It's the forester's equivalent of 'pick-your-own' rather than buying at a greengrocer—though with the difference that one can only take what has been selected! As the seller of standing trees the price offered is the money made; there are no additional expenses beyond preparing the parcel for sale in the first place.

Before measuring and preparing the parcel of timber, the fallen pines blocking the cross rides needed clearing. Five lay like 'pick-a-sticks' and sorting them out was as tense as the child's game and called for as much thought and care. Blown trees still hinged to the rootplate are exceedingly dangerous if cut in the wrong place. One's inclination is to cut it off neatly at the butt, now presented so conveniently by the prostrate tree. This can be fatal. Even for moderate sized pines there is half a ton of stem and branches pinning down and keeping bent the reluctantly wrenched-out roots. If cut through near the base, the whole weight is released suddenly like springing a trap. Instantly, the rootplate flips back into its hole, all too often hitting fearfully and hospitalising the prone, inexperienced operator as it does so. Sometimes the forces unleashed pinch the saw instead, gripping it vice-like as the stump being cut off vies for freedom with all that is weighing it down. Either way there is great danger starting at the bottom. Better is to work steadily, length by length, down the stem from the top, after first cutting off the side branches, or 'snedding' them as it is called. An ambulance should not then be needed as the tension between heavy tree and strained roots eases little by little.

Once the rides were clear the edge trees beside them and the main track were high pruned. This avoids damage by extraction equipment and reduces shading to help the rides and track stay dry. A couple of trees were felled at the bellmouth entrance to improve access and turning for lorries. The track and rides were swiped, and the gate's hinges oiled, but little else was needed to aid extraction in the wood and enhance its appeal to potential buyers of pine.

The main job in preparing trees for sale is to estimate accurately the quantity of timber on offer. Measuring trees is an inexact science. Counting the number for sale is easy—in our case the figure came to 2065—but accurately determining their volume, the usual measure of quantity used, is difficult. A tree's irregular shape and numerous appendages defy precise measurement of volume: one can hardly follow Archimedes and immerse the whole thing in water and measure the displacement. The first simplification is that branchwood and tops of trees are assessed separately, if at all, from the valuable part, the main trunk. But even the trunk is a problem in solid geometry and systems of approximation use the diameter of the tree at an internationally agreed position, called breast height, which is 4 feet 3 inches above ground. Until quite recently the diameter was measured in a special kind of inch called 'quarter-girth' which converted the girth of a tree, that is its circumference, to 'diameter' by dividing by four rather than pi. This was devised by Edwin Hoppus, a surveyor to the Corporation of London Assurance in the eighteenth century, and underestimates the true volume by about 20 per cent. Hoppus was happy with this discrepancy to allow for waste because of the bark and from sawing the trunk into a square balk before planks were cut off. Such 'hoppus measure', as it is called, is still used for selling large oak, ash, and beech and a merchant will talk of a tree as being so many 'hoppus'.

I have a 1790 edition of hoppus tables, grandly inscribed as *Mr Hoppus's measurer greatly enlarged and improved* which give tables of square measure and volumes for particular diameters. Poor Edwin Hoppus devotes much of the preface contending with the tables of others: Mr Darling's Carpenter's Rule made easy, and Mr Isaac Keay's Practical Measurer, which he says are the books most buyers and sellers of timber would rely on. Hoppus justifiably pillories Keay's rule as '... so far from the truth, that a greater fallacy could hardly be offered'. Keay's rule said that to calculate the cross-sectional area of any four-sided object one simply adds up the lengths of the four sides and divides the total by four. With several examples and much censure Hoppus roundly disproves this as humbug, as a moment's thought of how we are taught today will concur. How much timber was so erroneously sold we will never know.

The Forestry Commission's Tariff system, devised in the 1950s, was used to estimate the total volume of our 2065 pines. A sample of trees is measured for diameter at breast height, and a sub-sample of these is felled to estimate accurately their volume on the ground. The relationship between volume and diameter provides what is called the 'tariff number'. A table for each tariff number shows what the tree volume is for each diameter. For 2065 trees the system required measurement of diameter of every sixth tree and felling of every sixtieth to determine its volume and find out the average tariff number for the whole crop. I used a plastic girthing tape which marks off centimetres of diameter directly from the circumference. The work is much easier than this account might suggest, but it takes quite a time. One must get right up to each sample tree and carefully determine breast height by using a stick exactly 1.30 m long, the specified height in British forestry. A tape is run round the trunk at this point, making sure it is level, not twisted or askew, and not passing over any swelling. All such faults exaggerate the diameter. With pines a whorl of branches can often coincide with this measurement position and an average of the diameters above and below is taken. Sticky fingers from the resin are inevitable, though its aroma is rather pleasing. However, getting pine resin off one's hands requires neat washing up liquid or even Swarfega, and to remove from clothes, a visit to the dry cleaners. The special sub-sample of pine trees were felled using our small Stihl chainsaw. Two of the trees snagged in the branches of neighbouring ones and wouldn't fall to the ground unaided. Shaking, swinging from side to side, and finally levering the butt forward bit by bit using a pole like rowing with an oar, brought the obstinate trees to heel. Once on the ground

Practical Meaſuring

MADE EASY

To the MEANEST CAPACITY

BY A

NEW SET OF TABLES:

Which ſhew at SIGHT,

The *Solid* or *Superficial* Content (and *conſequently* the Value) of any *Piece* or *Quantity* of *ſquared* or *round* Timber, be it *Standing* or *Felled*, alſo of Stone, Board, Glaſs, *&c.* made Uſe of in the *Erecting* or *Repairing* of any Building, *&c.*

Contrived to anſwer all the Occaſions of *Gentlemen* and *Artificers*, far beyond any Thing yet extant: The *Contents* being given in *Feet*, *Inches*, and *Twelfth Parts* of an Inch.

WITH A

PREFACE,

Shewing the *Excellence* of this *New Method* of Meaſuring, and *Demonſtrating*, that whoever ventures to rely upon thoſe OBSOLETE Tables and Directions Publiſhed by ISAAC KEAY, is liable to be deceived (*in common Caſes*) 10s in the *Pound.*

By E. HOPPUS,

Surveyor to the CORPORATION of the LONDON ASSURANCE.

The TWELFTH EDITION

LONDON 1790.

they were then measured for length of stem from the base to the point where it narrows to 7 cm diameter, the old 3 inch cut-off point. At the mid-point along this length the diameter was measured for calculation of the tree's stem volume. All measurements were rounded down as required by the Tariff system.

Some of the Corsican pine in the wood had cankers on the trunk caused by the fungus *Cremenula soraria*. Crusty ulcerations had erupted at intervals along the stems of infected trees. Two of the trees felled for tariffing and some of the windblown ones had these bark lesions, and I took the opportunity to cut through several to investigate what damage was done to the timber. There was no sign of stain or decay in the wood beneath and, apart from stem distortion, the ugly cankers were more an alarming disfigurement than a serious defect. The presence of the disease was noted with the details gathered about numbers of trees and total volume.

All this work took a whole day. By evening I knew we had 2065 pine trees, with an average diameter of 21.8 cm (8.5 in.) and a volume of 636 cubic metres, equivalent to this number of tons or around 20 000 cubic feet. These figures for diameter and volume are, of course, still only an estimate and a buyer must inspect the stand to satisfy himself that they are about right. From time to time disputes arise and the estimate based on the Tariff system is challenged. On such occasions, the Forestry Commission's mensuration experts are called upon to arbitrate. Janet Methley runs this unit and she usually finds either that the sampling wasn't done quite right or that the buyer left timber behind by not cutting trees off properly at ground level, or by failing to harvest all the wood available. The Tariff method itself has stood the test of time and generally can be relied upon to be accurate to within about 10 per cent—trees are really a very awkward shape!

In late March the felling licence application to cut the pines was submitted to the Forestry Commission. The licence came in early July and describes the operation that is authorised, silvicultural thinning in our case, and shows how the trees are marked, the species, the area of ground, the number of trees, and the estimate of the total volume for which permission to fell is granted. Before issuing the licence the Forestry Commission consults with several organisations—statutory consultees—to see if there are objections or special conditions which should be attached. Usually this only affects clear fellings, and then only where a site is of special wildlife conservation interest or is an important landscape feature; licence applications to thin normally go through quite quickly. The licence is valid for two years by which time all the work must be completed. The next task was to find a buyer.

I prepared a schedule of the particulars giving all the above information. Details were included about location, access, timing, and special conditions such as how the debris from lop and top should be disposed of. Potential buyers could ask to see the volume estimate calculations. Standard conditions would also apply covering late payment, not harming wildlife, restoring the main track to good condition, and so on. These particulars, with a map of the wood showing the tracks and where cut timber could be safely stacked, were sent to 13 potential buyers. These were selected from lists of timber merchants and purchasers which the Forestry Commission produces for each county.

The particulars inviting tenders were posted on 6 July. The actual number sent, 13, was not intentional, simply that we were keen to get a cross-section of potential interest. A fortnight later the first reply arrived. I looked at the unopened letter—the envelope advertised a timber company, Tilhill Forestry—and turned it over slowly, full of anticipation. Tilhill were not interested and made no offer.

My deflation was a surprise. Only then did I realise what expectations had been building up as the trees had been carefully prepared for sale, a volume estimate calculated, a licence applied for, particulars drawn up, potential buyers identified, and invitations to tender sent off. It was a matter of professional pride to do the job well: John was relying on me, I was the one supposed to know how to sell trees. What if no one wanted them, how would John realise his investment, would they stand for a few more years? Disappointment is a passing emotion, soon replaced by rationalising or, usually more truthfully, making up excuses. There was a glut of timber on the market because of the storm, and we were a new supplier. Or, was it that the total quantity of timber on offer was a bit too small for Tilhill, Britain's largest private forestry company? Perhaps the proportion of good-sized logs to small ones was not attractive enough since ones suitable for the sawmill only made up half the parcel? Or was it simply the recession, or that July is the beginning of the summer holidays? The list went on and on. Perhaps trees are just like everything else: it's always a bad time to sell when you want to, whether it's a car, or a house or simply outgrown children's toys. And now trees.

The next reply came two days later. It was a second rejection, and this time it was even without a hopeful remark, which at least Tilhill's letter had contained, indicating that they might be interested in other timber we might have to offer in the future.

Further contemplation about the sale was, for a time, set aside when on 4 August Margaret and I travelled to Canada for an international forestry conference. I was presenting a paper about the long-term productivity of tree plantations, based partly on our many visits to Swaziland to research the subject. On returning home a fortnight later, two more letters had arrived. Opening mail after being away is one of the after-pleasures of a holiday or break, bills and the Inland Revenue excepted, but the two letters of rejection already received created distinctly mixed feelings about the new ones. They were consigned to the bottom of the pile beneath magazines, a Tear Fund request for advice on social forestry among Afghan refugees in Pakistan, junk mail promoting life insurance, double-glazing, cut-price Caribbean cruises and other unsolicited bumph destined for the bin, a letter from missionary friends in Ethiopia, and the National Trust's subscription renewal reminder. The pile was finally exhausted and the two letters remained; no longer could their opening be deferred. There is a finality to opening. Hope is instantly consummated or dashed. Like letters with exam results, or when hoping for some award, hope is sustained to the very point of opening, but no further. Slowly and deliberately the envelopes were slit and the letters extracted, unfolded, and read: the first contained an offer, and so did the second.

The tenders themselves differed by over 50 per cent with the poorer one a distinctly disappointing bid. During the last week, before the closing date of 24 August, no further offers or letters of rejection appeared on the doormat. I began preparing the contract of sale for the one offer which John and I were willing to accept. On the day before the closing date our son Jon took a call at home from another company who wanted to buy. They phoned the office to find out if there was still time to put in a bid. This duly arrived, offering a price comparable with the better of the two existing offers. The company was Mendip Forestry Ltd who had worked in the wood before when cutting the large timber at the bottom, and whom I knew by reputation. We decided to accept their offer.

Of the 13 invitations sent out, we received five replies and three firm offers. The market-place had worked.

The contract documents were signed and sent off to Mendip Forestry with a letter accepting their bid. We requested evidence of the normal third party insurance cover and asked for the first of just two instalments to be paid by the end of October when they returned the signed contract. Thus Mendip Forestry became the owner of 2065 pine trees in our wood, and they had acquired the right to harvest them at any time over the next 18 months. This interesting transaction also gives the buyer all new wood put on while the trees are unfelled and continue to grow which, for the pines, would amount to several tons per year. In the contract Mendips were also free to take any of the blown or leaning pine and Douglas fir felled by January's storm. In the event many of the windblown pine were severely blue-stained by a fungus which invades the timber, making it unattractive and almost valueless. While a pine tree is still alive it successfully repels blue stain infection, and I hoped that the leaning ones and even those on the ground still with live branches would remain free. None of these trees had been counted in the original estimate, and were extras for Mendips to take if they wished.

The price we got for the pine trees was rather more than £13 per cubic metre, or about 50p a cubic foot. Since about three of the trees would constitute a cubic metre of timber we can say we got about £4 per tree; not very much after 34 years' growth. In fact the figure of £13 per cubic metre was close to the average obtained by the Forestry Commission in 1990 and 1991 for trees of a similar size. It publishes this information each year and at the time small conifer trees were fetching about £10 per cubic metre and large ones a disappointing £20 or so, largely because there was so much timber on the market following the storm. Average tree size is not the only consideration. Ease of access to and within a forest, availability of stacking areas or loading bays, whether the stand of trees is being completely felled or just thinned (as in our case), and the quality of the trees and freedom from defect are all taken into account when being valued. Average prices are just a guide.

We were reasonably pleased with what we got. In total it came close to the purchase price paid for the wood five-and-half years before. Of course inflation had intervened, but we still owned a 22-acre wood, albeit no longer bearing its most valuable asset of over 600 tons of pine. And, unlike capital gains on British Gas shares, income from the timber sale was entirely free of tax thanks to the Chancellor's announcement on budget day, 15 March 1988, taking forestry out of the tax environment. Unfortunately this is not the loophole it appears since, even prior to 1988, income from timber sales was hardly taxed, being treated as if the land on which the trees grew had been let as unimproved grazing. One would pay tax as if earnings were £20 or £30 per hectare, which unimproved land might make, whereas one's timber crop at maturity could well sell for £5000 per hectare or more. Also, of course, as an investment starting from scratch, this return is paltry because of the long delay, half a lifetime or more, between planting and final felling. Rarely will one make a return of more than one or 2 per cent interest in real terms. As the forestry investment companies advocate, it is best to buy a well established woodland with perhaps 10 or 20 years to go before final felling. This is what we did and the result so far bears out this investment advice. But if a forester and solicitor together couldn't make a go of forestry, it would be a poor show indeed.

The next significant income will come from felling the patches of Douglas fir and then from occasional thinnings of beech, though both are small beer compared with the pines. The pines sold to Mendip Forestry in 1990 had served their purpose. They helped stock the ground and helped the beech to grow, and they enabled the owners to obtain a return sooner than that yielded by a woodland of pure beech. Their removal also helps in another important way: the remaining beech have much more space in which to grow.

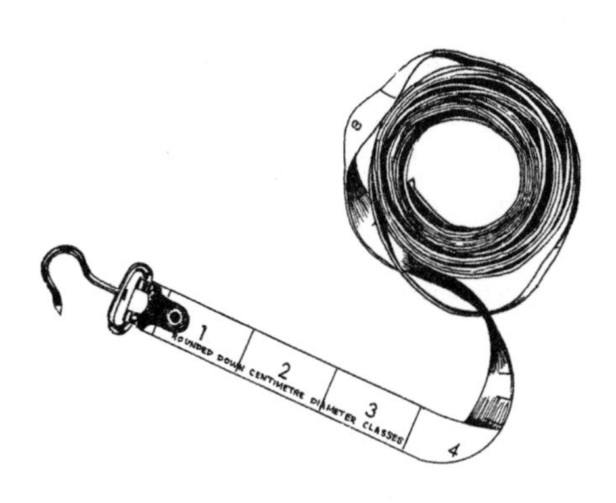

11

Harvesting the pine

More than a year passed before Mendip Forestry came to cut their pine trees. I had not expected such a long wait, though it was entirely within their right. Indeed, I probably pestered them a bit, wanting to know when they would start. It is not always easy for a forestry contractor to plan his work precisely owing to our fickle weather, unforeseen machinery problems, clogged log yards at sawmills, and the like; of course, there was also a lot to clear up after the great storm. In the event Roger Austin, Mendip's Director, phoned in early April 1992 to say that they would be shortly starting to cut. By the time I next visited the wood, work was well under way. Access had been gained by using the keeper's key and his lock. Subsequently ours was hung for them on the back of the entrance sign. In anticipation two extra keys had been cut to hand over when work started. This is particularly helpful for lorry drivers arriving at all sorts of hours to pick up cut timber. This can continue for months after the main job in the wood of felling and extracting the trees is finished, and the driver needs his own means of access.

Mendips employed a contractor from Wiltshire, the brothers Terry and Gary Price of Trowbridge, to cut the pine. They arrived on 6 April and were gone 12 days later, making the contract deadline with ten days to spare. Their speed of working through over 2000 trees and cutting more than 600 tons of wood was remarkable for a two-man team. They brought in two machines, one a harvester so new that it was barely past the evaluation stage in

British forestry. Its bright yellow paint was unscratched and it showed off all the rubber piping possessed of modern hydraulics like some mechanical patient nourished by intravenous drips. Word of its presence got around and the local contractor, Alan and Henry Robinson of Deane from just down the road, came to the wood to see it at work.

The harvester, a Hymac, was a tracked vehicle with a long jointed boom with the words 'Forest Blazer' stencilled proud for all to see against the yellow. A special cutting or felling head, vividly jade green and attended by black and yellow pipes and hoses, was mounted at the far end. The operator, from the comfort of his cab, would manoeuvre the cutting head, hovering like a praying mantis at the end of the long arm, up to a tree. Once in place, two powerful steel callipers, about 4 feet apart, quickly extended to grip the trunk. As the callipers locked home, a saw at the base of the cutting head severed in seconds the tightly held tree from off its stump. The cutting head, still gripping the tree, then lifted the whole thing, trunk, crown, and all, as if to toss the caber, and proceeded to work along the trunk shearing off the branches. As it did so, the saw whined into action at set intervals and cut off logs of the desired length. Next to where the logs dropped from their slicing machine, the branches rapidly accumulated in a fan-shaped pile. After the topmost log was cut, the treetop was left where it fell like some huge discarded bottle brush. The whole job of felling, debranching, and cutting up into logs took just a minute or two per tree. The use of this impressive equipment was only possible because both rows of pine were being cut and there was adequate room for operating. Remarkably little damage was done to the neighbouring beech.

As the harvester worked along the double rows of pine, piles of neatly stacked logs were deposited as if preparing for some giant under-seven's relay race with logs rather than bean bags spaced out. The one job this machine could not do was gather up its product and stack it at the entrance to the wood to await collection by lorry. This fell to the second member of the team who operated a smaller vehicle, called a forwarder, which can manoeuvre in narrow spaces, being hinged between the axles. It has its own loading arm and grapple to lift bundles of five or six logs at a time. The forwarder, with all the busyness of a worker bee attending its queen, scavenged for the harvester's output to keep the ground clear to allow work to continue uncluttered. Once loaded

with 6 or 7 tons of logs, the forwarder scampered off up to the entrance to unload its burden and pile up the logs with its antenna-like grapple arm. Emptied, it then hurried back to attend again the productive harvester.

The harvesting was efficient, and impressive to behold. On a tiny scale, the smooth running of a sequence of connected operations all to achieve one end reminded one of reports of the building of the Panama canal. This channel from the Atlantic to the Pacific was for long held up by the enormity of excavation through the Culebra cut. The problem was not excavation itself, but how to collect, take away, and dispose of what was dug out at exactly the speed the excavating machines were working. And the amount to be disposed of was 91 million cubic yards, or rather more than 25 times the volume of the great pyramid of Cheops. Gaillard, an army engineer, ran the operation in those confident Edwardian days, using steam shovels and steam railways whose tracks had to be moved nightly to service the excavators. So efficient and impressive was the operation that reporters and sightseers flocked from the States in their thousands to brave the torpid equatorial heat, the perpetual, enervating humidity, the risk of yellow fever, and the much-dreaded Chagres sickness—a particularly virulent form of malaria—just to marvel at the spectacle. Contemporary descriptions abound with superlatives. The metaphors of beehive or ant colony were pressed into service to convey the incessant activity, appearing random and undirected from the high vantage point overlooking the cut, but all brilliantly marshalled and inexorably cutting a path between the oceans. For many it remains the greatest civil engineering feat of the twentieth century.

It is perhaps incongruous to compare construction of the Panama canal with 12 days' harvesting of pine trees, especially since Roger Austin of Mendips later told me about the particular harvester's unreliability. Soon after working in the wood, the company replaced it with an even more sophisticated machine, with an eight-wheeled rather than track-based unit, which they imported from Finland. Nevertheless, I marvelled at the Price brothers' two-man operation in the wood. All the more so because several days were very wet, though not tropical in warmth nor disease-ridden beyond the remote chance of tick fever! Up to late March the winter had been unusually dry and ground conditions ideal for harvesting trees. The beginning of the cutting coincided with the beginning of a prolonged damp spell, leading to gouging and rutting of both the main track and the two cross rides. Much of the splendid display of primroses opposite the grey water tank was obliterated by the passing and repassing of the tracked Hymac and the industrious forwarder. In the years since, the flowers have gradually recovered though, even now, leaves are small and primrose blooms weak. Regret over this damage is ungenerous because the contractors were exceptionally careful and rarely scuffed or barked any of the beech trees. The only other consequence of the rutting was uneven tracks and rides. Once work had finished the highest ridges of congealed mud were crudely and stickily raked into the adjacent rut to smooth the roughest parts. This irksome job was mostly needed next to the water tank where the cross rides intersect the main track and where the two machines had continually turned and manoeuvred and gone back and forth. A small JCB for a couple of hours would have done the job. Instead, odd hours of labour, much rain, and growth of vegetation have weathered away the undulations.

Compared with old systems of cutting and extracting trees, the wood as a whole was hardly damaged. Even today much of harvesting work is still done by skilled tree fellers using chainsaws, with a tractor and drum winch to drag out the logs or whole trees to the loading area. Dragging logs behind a tractor, called 'skidding' or 'tushing', badly cuts up the soil surface in damp weather. Even worse is the ease with which the long tail of logs behind the tractor almost unavoidably rubs or snags on the trees left to grow on. Lumps of bark are scraped from the trunk and root spurs as the tushed log abrades all along its length like some giant file. Horses and mules are often more careful at this particular job than tractor drivers, and the lighter tread compared with a tractor's wheel is also less damaging to the ground. There is currently a resurgence in their use, especially for thinning out trees when space is severely limited. On all counts they are more environmentally friendly than machines, but only where wage rates are low, such as in many tropical countries, is the extensive use of draught animals really feasible.

The pine logs were stacked at the entrance and separated into three sizes. The largest, the sawlogs, were kept to the north side and made two huge stacks 10 feet high and perhaps 80 feet long. Medium-sized logs, 'bars' to the trade, were stacked on one side of the turning bay, and the smallest logs, cut from small trees and the tops of large ones and suitable for fenceposts, chipping, or pulpwood, were piled up on the other side. The few Douglas fir logs obtained from blown or leaning trees were stacked next to the gate. It was important that all the timber was assembled around the entrance since only 25 yards of track and the turning bay are properly formed and with stone metalling which is capable of supporting loaded timber lorries. Farther into the wood, the track is soft and lorries would soon get bogged down whenever it was wet. Even so, there is little room, and on at least one occasion a lorry must have got into difficulties as the entrance fence took a battering. By the end of April the entrance looked like a log yard with the enormous piles of timber having, as it were, been attracted to this central point at the top of the wood like iron filings to a magnet.

Hauling of large logs to a mill in the Midlands started almost as soon as they were cut. The long, 10-feet-high stack at the entrance was continually depleted and replenished as lorries hauled and the ever-conscientious forwarder gathered in the trail of the

harvester. Even so the lorry haulage didn't keep up and the last logs didn't go until June, some two months after felling finished. This was none too soon to ensure that the ingress of blue stain fungus had not begun to discolour the wood. For pine one normally aims at no more than six weeks elapsing between felling and sawing up in the mill. Some of the smaller pine logs were never taken and their rotting pile remained for several years. And, for almost a year, a full load of more valuable Douglas fir patiently waited, but these suddenly went and the chain on the gate suddenly shortened. It was, doubtless, a very late pick up by Mendips and not theft; the logs weren't that valuable.

Only about 6 or 7 tons of pine logs, out of 600 or more, were left behind, barely a lorry load and probably the reason they were not bothered with. They are now ours and have been shared with two generations of blackbirds who have enjoyed nesting rights in a crevice where stacking was untidy. The only possible use for the logs might be to chip them for mulch. Wood chips, unlike bark, make poor mulch unless old and long weathered, because they take more from the soil than they contribute, depleting it of

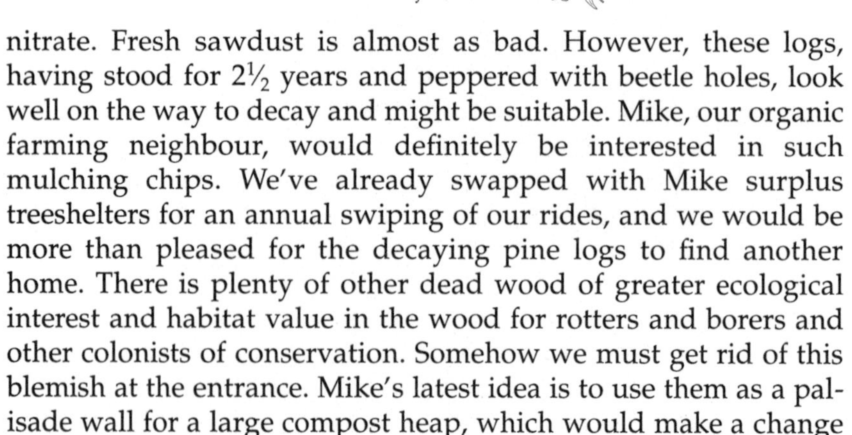

nitrate. Fresh sawdust is almost as bad. However, these logs, having stood for 2½ years and peppered with beetle holes, look well on the way to decay and might be suitable. Mike, our organic farming neighbour, would definitely be interested in such mulching chips. We've already swapped with Mike surplus treeshelters for an annual swiping of our rides, and we would be more than pleased for the decaying pine logs to find another home. There is plenty of other dead wood of greater ecological interest and habitat value in the wood for rotters and borers and other colonists of conservation. Somehow we must get rid of this blemish at the entrance. Mike's latest idea is to use them as a palisade wall for a large compost heap, which would make a change from old railway sleepers.

One other matter during harvesting of the pine remained unresolved. Two or three days before he finished, the Hymac operator enquired if there was any more pine. At first I misunderstood, thinking he was asking about opportunities for additional contracts. What he meant was, were there any more pine trees in our wood which were part of the present job? He could see only a few more to cut and felt short of 50 or 60 tons, judging from the number of loads the forwarder had hustled to the gate. I told him there were no other trees, so had I overestimated the quantity? The estimate was based on the accepted Tariff system, but even the book says it is only accurate to within 10 per cent. And trees are still such awkward shapes! Other reasons for the discrepancy were that the Hymac appeared to leave highish, even proud, stumps and that a number of small piles of cut pine logs were never scooped up by the forwarder. But all of these considerations together would not account for the whole shortfall. The contractor may be right and the estimate could have been on the high side, always assuming his own rough load-by-load estimate was itself close to the mark.

In the years since the pine were cut, another untidy pile of logs has appeared near the entrance beneath even more untidy sheets of plastic. This pile is of our own making and is a mix of beech and sycamore cut for firewood. Supplies from the patch of sycamore coppice in Taid's wood ran out after five years and we cast around for what else we could cut. At the top of the wood next to the lane is about three acres of what had been called low-grade broadleaves though it was really a mixture of scattered Douglas fir trees, some medium-sized oak standards, and planted trees of sycamore and beech. It was not thinned in the Forestry Commission's time and the only treatment since has been a light thinning for firewood which Martin Wagner's Smallwood Services did just after we bought the wood. It needed attention, both owing to this relative neglect and because patches of beech trees were dying in small groups from outbreaks of beech bark disease, compounded by lime chlorosis and squirrel damage. These trees became the new source of firewood.

Both Jon and Stephen, and more recently Ben, are part of this sideline of firewood merchanting. As stressed before, trees would be cut one winter for the following winter's firewood supply. I usually did the felling and cutting to length of dead or dying trees with the small chainsaw and the boys carried out the logs to the rideside stack. It is heavy work and we confined the cutting and tidying up to within about 40 yards of the stack. Unseasoned wood full of sap, even from recently dead trees, is heavy and, not infrequently, carrying the logs either became a two-man job or they were tediously and tiringly pitched end over end to progress one log length's distance at each exertion. It is back-breaking and quite the antithesis of the scampering forwarder. Most years about 1 or 2 tons of firewood are cut and stacked as short logs of about 4-foot lengths. Generally it takes us a day to cut, extract, and stack this amount.

The firewood was not sold in 4-foot lengths to our five regular customers. We cut to order from the best seasoned stack, selling ready-split logs suitable for a normal grate. A sawing horse was created out of two pairs of old chestnut stakes driven in at an angle and lashed together with mild steel wire at the cross over point. They looked rather like the beginning of two squat Indian tepees a couple of feet apart. A log would be placed on this makeshift, but remarkably durable sawing horse, and cut into 7- or 8-inch lengths. For most of the time lifting logs on to the horse and then cutting them to length was my job, and Jon and Stephen would do the splitting with either a hand or felling axe. Orange polypropylene string bags made by Tildenet, of the type used for carrots and cabbages, were filled to overflowing and tied off at the top. These bags are tough and have the advantage of allowing good air circulation to dry the wood further, and people can see exactly what they are getting. A full bag of seasoned wood weighs about ten kilograms, a bit over 20 pounds, and is a far more substantial offering than those found at garages and garden centres.

And, our price of £3 delivered is far better value. Indeed, at a garage near Basingstoke, a bag containing just six small logs was seen priced at £3.99, and that was without home delivery. Our delivery service is, however, the one weak link. With the back seats down the car will hold about 15 bags and Jon can get about 12 in his Cavalier. Family cars were not built for transporting firewood, and if we ever enter the trade seriously both a van and a log splitter would be early purchases.

The handful of customers to whom firewood is delivered each year like their logs cut ready for the grate. Also, the bags seem about the right size for convenient stacking in a garage or bringing into the lounge for the fireside basket. Two of our bags are sufficient to fill a standard dustbin. All that we cut gets sold. And, since 1993, we have a fireplace in our own home and so can market test the product too! Also since 1993, our youngest son and his friend Kenny have joined in firewood production, Jon having moved on to the more cerebral concerns of economics and business finance at Brunel University. He perhaps learned a little from our tiny firewood business, which still amounts to less than £100 per year: that there are many stages of added value such as hard work, as dead tree is converted to dry log suitable for the grate and worthy of payment.

I must conclude by returning to Terry Price, the Hymac operator cutting the pines, because he passed on an important piece of intelligence. During his 12 days in the wood cutting the pine, he reported destroying several squirrel dreys (the large nests squirrels build in winter like a rook's nest only lower down and always tight up to the main trunk) and disturbed not a few squirrels. This was despite the attempts to control grey squirrel numbers because of the damage they do. By mid-summer the implications of this report were there for all to see. More than 100 beech trees were debarked at the base, several quite severely, because of stripping and peeling by squirrels. The contractors had carefully avoided doing any such damage, but grey squirrels struck instead. It was upsetting: the greys had come.

12

The coming of the greys

Grey squirrels tore bark from many of the beech trees the contractors had been so careful to protect. Just two months after the last pine was cut the first signs of this peeling were evident. Fresh, pale cream scars appeared at the base of several trees, their light colour blatant and taunting. Flakes of bark were strewn at the scene. I knew the culprit as every forester working in Britain's broadleaved woodlands does: the grey squirrel. This outbreak of stripping was much the worst in our seven years of ownership. But why had it flared up in the year the pine were thinned out, or was it just coincidence?

The grey squirrel is a recent arrival from America. Although there is a tantalising account of grey squirrels breeding in Montgomeryshire in 1828 and a report of one being shot in Denbigh at about this time, the first authentic introduction was a release of a pair at Henbury Park in Cheshire in 1876. No doubt the intention was to keep the new curiosity captive. This was impossible from the start and with a further 30 or so deliberate introductions over the next 50 years, grey squirrels soon began to spread. Our climate was congenial, the food of acorns and beechnuts ideal, and the competition weak. And, unlike home in New England, there were not even any predators. In one hundred years they have spread to most parts, though not yet north into the Scottish highlands. They are also, mercifully, still absent from the Isle of Wight, Brownsea island, and other isolated enclaves and, remarkably, none occurs in Europe except for one small colony inhabiting fields of maize and poplar trees in Italy's Po valley.

Grey squirrels strip or peel bark from trees such as beech, oak, and sycamore. Very young trees are not harmed, but once the bark has some thickness and the trees are anything from ten to about 40 years old severe damage can occur. Any part of the stem or larger branches may suffer, though root spurs can be a favourite. If, as often happens, the squirrel peels bark right round the stem, every part of the tree above soon dies. Basal ring-barking is quite common and trees are consigned to a premature end in just a few minutes' stripping. Stripping is a summer occupation of squirrels mainly in the months of May, June, and July, and is done for no very apparent reason. Little nourishment is forthcoming, the bark itself is chipped off by the squirrel's front incisors, and the flakes simply dropped; they are certainly not chewed or eaten. It is also known that not all grey squirrels peel bark; and it tends to occur when numbers are high and many young males present in the population. Stripping may be a displacement activity reflecting stress in the squirrel pecking, or rather peeling, order.

The Timber Growers Association, who represent many of Britain's private woodland owners, count squirrel damage as the single greatest threat to our lowland broadleaved woods. Numerous stands of potentially fine timber have been spoiled beyond repair by this menace of beech and oak.

Serious attempts at control began before the last war with the founding of the National Anti-Grey Squirrel Campaign in 1931. A year or two later the Ministry of Agriculture also began urging their control. But it is the post-war bounty scheme that is the most celebrated and probably the most ineffective attempt at control. So devastating was damage from grey squirrels, that in 1953 the government of the day offered the bounty of a shilling for every squirrel tail handed in to a Ministry of Agriculture office. At that time those with a shotgun licence could come away from the same office with a free box of 12-bore cartridges for killing vermin, at least that was the Ministry's intention, with each cartridge having 'pest control' imprinted on the case. Neither the gun's owner nor certainly the quarry would ever reveal if sometimes this was not the only use! And so grey squirrels were added to the list of vermin. The bounty was doubled to two shillings per tail in 1956, but the scheme only lasted five years, not because grey squirrels had been eliminated, but because it was realised that paying for the tail didn't necessarily mean the animal was dead nor its reproductive abilities impaired. The size of the task was enormous, wholly untargetted, and cost the taxpayer £80 000 in all, or about £2 million in today's terms. But the government didn't forget about grey squirrels. The subject even surfaced in Macmillan's cabinet, and throughout 1960 and 1961 the Prime Minister personally fired off minutes to ministers and departmental secretaries enquiring what was being done about them. He believed that withdrawal of the bounty had caused their resurgence and dismissed the advice of scientists, who suggested that populations were related to weather patterns and food supplies, by saying 'What do scientists know? What I say is true and known to every keeper in Britain.' I haven't verified the source of the scientists' advice which caused prime ministerial derision, but doubtless it emanated from my research station at Alice Holt Lodge.

That a pest should attract such attention isn't altogether surprising when they are so pervasive and damage is so widespread. Doubtless, too, many in the Cabinet, not to mention the Upper

House, owned woods which were suffering. The sight of vigorous, healthy oak or beech, perhaps 30 or 40 feet tall and well past all the dangers of initial establishment, stripped of great patches of bark as if gnawed by a giant caterpillar, is deeply depressing: 30 years of care destroyed in 30 minutes. One is dismayed. Trees are killed or maimed in their prime.

Damage caused by grey squirrels is not confined to stripping bark. In the spring they eat buds and unfolding shoots and have occasionally led to shoot dieback and death of twigs over a whole crown. Squirrels are certainly known to raid birds' nests for their eggs, chaffinch being their first preference, and they will even prey on young chicks. To what extent these activities seriously harm wildlife is unclear and needs investigating.

However, the grey squirrel is certainly not a universal evil; indeed, quite the contrary. For many people seeing grey squirrels in the garden brings the countryside to their home. Their attractive bushy tails are endearingly cocked up as they nibble, bit by bit, on a nut held delicately between the front paws. Large, warm eyes are ever alert. Their agility, curiosity, and inquisitiveness are appealing and, in large part, make up for frightening off the less adventurous and more timid of garden visitors and devouring everything on the bird table. As many know to their frustration, feeding birds and enjoying grey squirrels are virtually incompatible activities. Nevertheless, the grey squirrel entertains and endears, and is fun in the garden.

Domestication of grey squirrels is not confined to their conquest of suburban gardens. Taking up residence in lofts is another manifestation much on the increase. They get in through a hole or gap in the eaves and have taken to chewing electric cables with literally shocking results. Harry Pepper, the Forestry Commission's squirrel expert, has, on more than one occasion, advised builders on how to squirrel-proof the loft space. Unfortunately, getting rid of squirrels which find loft insulation or even the cavity between walls irresistible, is the job of the local pest control officer: using methods of control permitted in the forest is illegal in the home.

The grey squirrels' other notoriety is the steady vanquishing of their red cousins in Britain. The grey is a recent introduction now at home in these islands. It is better adapted than our native red to exploiting the large seeds of acorns, beechnuts, and hazelnuts, but it is not so happy in coniferous forest. It is in such forests that the red has held out. Indeed, this century's large plantations of spruce, and especially pine, have probably saved the red squirrel possibly just long enough to allow research to devise a way of selectively controlling only greys where both species are present. The 1973 Grey Squirrel Control Order does not permit this at the moment, but a research breakthrough may change this.

Control of grey squirrels is possible in several ways. Trapping or poisoning with warfarin as is used for rats, though in a different formulation, are mostly resorted to. Other methods which keepers use, such as drey poking and shooting, add to wintertime activity and may be good sport but have little effect on squirrel numbers since during the day most animals are away from the drey feeding. And a destroyed drey is often rebuilt, or the squirrels disperse to a comfortable hole or a less troubled area nearby. Trapping uses a wire cage with a door which springs shut when the squirrel presses a treadle. The animal is enticed into the cage by grains of whole maize, though grey squirrels need very little encouraging to explore tunnels or follow challenging passageways. Once traps are set the law insists on three things: the trap must be checked at least once a day; any grey squirrel caught must be killed in a humane way; and no grey squirrel may be released back into the wild.

The other method of controlling grey squirrels with warfarin became legal in 1973 in all counties where red squirrels were no longer found. Sadly the wood has no red squirrels and neither does anywhere else in Hampshire; the greys displaced them long ago. (Indeed, the nearest spot where the native red squirrel still survives is the Isle of Wight, simply because the grey has not got there.) Instead of traps the method uses specially designed hoppers which have a heavy flap door in the tunnel which deters all except the aggressive grey squirrel, whose inquisitiveness brooks no obstacle. All other animals are denied access to the bait.

As mentioned, the need to control grey squirrels is limited to only some stands of trees in some broadleaved woodlands. Beech, oak, and sycamore are most at risk and then principally between the ages of about 10 and 40 years. For most of a tree's life, when very young, and for the long years as it grows to maturity aged 80 or 100 years, or even 150 years in the case of oak, grey squirrel damage either does not occur or is not lethal or debilitating. When we bought our wood the main tree species, apart from the pine, was beech along with a scattering of sycamore, and their age was 27 years: the trees were a vulnerable type and right in the middle of the highly vulnerable period. The wood was a likely candidate for the attentions of grey squirrels.

Coping with the grey squirrel problem has not been wholly successful. Not all damage has been prevented. In 1992, after the pine had been felled, extensive bark peeling occurred, and it was even worse in 1993. There seem to be several reasons for this, including in part circumstances beyond our control as I shall relate, beginning with what happened a couple of years before. In 1990 there were increasing signs that squirrels were about. Indeed, during one particular visit I disturbed a plump squirrel feeding contentedly just inside the adjoining wood. Normally squirrels are quick to hide—they can race as fast as most people at speeds up to 18 m.p.h.—or freeze on sight or sound of people, but this was the first time one had been so obvious, clearly feeling at home, and perhaps suggesting that squirrels were beginning to use the wood as a more permanent residence. At the time I thought nothing of it and damage that year was light.

Later that year the new keeper, who had taken over from Alistair when the estate changed hands, began active feeding of pheasants in the winter. In September half a dozen feed dispensers, beach ball in size and UFO-like, arrived and landed at intervals along the rides to encourage pheasants into the wood and keep them plump in the winter months and, presumably, for the table too. Undoubtedly, such feeding keeps pheasant numbers up and improves the sporting. Unfortunately, it also provides limitless food for other aliens such as grey squirrels and doubtless for rats and other woodland inhabitants too. The pheasant feeding dispensers disappeared in late January as the shooting season ended, but they had been in place just at the right time to keep the squirrels well fed and greatly improve their chances of surviving the cold damp winter months. This extra food had come on top of the first significant mast year for beech in the wood. Although only 34 years old they produced quite a crop of beechnuts. Both oak and beech only begin to set seed in any quantity from about this age and then continue to do so at longish intervals of five to ten years. The winter of 1990 had been good for the grey squirrels.

By May 1991 there were several dreys and definite signs of increasing numbers in the adjacent, much older woodland, not itself vulnerable, though ideal as an holding area. In the autumn the bright aluminium, extra-terrestrial feeding dispensers returned to fatten the birds and featherbed the squirrels over another winter. The combined effects were revealed during the harvesting of the pine in early April 1992. As they cut the pine the contractors reported destroying several dreys from which very much alive squirrels escaped.

Over the next weeks squirrels were obviously in the wood, and probably in quite high numbers. In late June signs of bark stripping appeared and quickly spread to perhaps 150 of the best and most vigorous beech. Damage was worst either side of the cross ride where the previous winter's pheasant feeding had taken place. Something else was also evident. Throughout the wood, wherever the freshly cut tops of the pines lay, half-eaten pine cones were to be seen, each neatly chiselled away to the core. The squirrels had feasted on the crop of cones brought to the ground when the pines were felled. The cones were full of nutritious seeds. No wonder their numbers were up, as the stripping damage was to reveal. Good winter feeding, followed by a mild, dry spring, topped up with a feast of cones must have confirmed our wood as suitable for a base.

In Autumn 1992 a new keeper introduced a more down-to-earth and even more generous way of feeding wheat to pheasants. Six 44-gallon drums with specially cut pouches for the grain were installed to allow continuous feeding. The winter was not hard and a large number of grey squirrels probably survived to produce large litters in the spring. During the winter I tried to deter future damage from some of the best beech by pruning branches so that the squirrel was denied an easy perch, and by piling brushwood around the base as a kind of obstacle course. By June, however, bark peeling and stripping of numerous beech and sycamore indicated that I had not succeeded. Piling of brushwood did appear to be a partial deterrent, but 1993 was the worst year for damage so far. Indeed by August several hundred trees had suffered varying amounts of de-barking. Some had even been completely ring-barked and were dying: fine, 35-year-old beeches, cut off in the prime of life. Undoubtedly the previous year's thinning out of the pine was partly to blame. The extra space this gave the beech stimulated their growth and the bark of such invigorated trees is known to be liked by grey squirrels keen, it seems, to tear them off a strip, especially, almost maliciously, going for the very best ones.

In the autumn the keeper kindly agreed to delay deployment of the pheasant feeding drums for several weeks, but he had shooting to consider and we were not in a position to insist. Battle was resumed in the next spring with renewed vigilance; the wood simply could not tolerate continuing the damage levels of 1993. Winter and spring of 1993/94 were very wet, though not particularly cold, weather which was unlikely to affect squirrel numbers much. However, such weather did promote lush tree growth in April and May across much of England, and soon reports began arriving at our research station of devastation by grey squirrels. The ease of peeling or stripping affects the amount of damage, and the good growth following the wet spring of 1994 made bark very 'strippable'. Trees, indeed whole stands of trees, were reported with bark peeled off from top to bottom and the forest floor carpeted with chips and curls of bark that crunched underfoot. Such intelligence strengthened our resolve and by July, although some fresh peeling was sustained on about forty trees, it was clear that the scale for us was much less than the previous year; a contrast with elsewhere in England.

We have won some battles, lost several skirmishes, but are still in the thick of hostilities, and will be for a few years yet. Happily, soon for us squirrel wars will be no more. By the time beech, oak, and sycamore are 40 years or older they steadily become less vulnerable as thickening bark becomes less strippable. Some damage may continue on root spurs and branches, but this is of far less concern. The wood will be past the most vulnerable period. Truce will be declared, hostilities will cease, and the greys themselves will be left in peace. No longer need this high-spirited arboreal member of God's creation be troubled.

In the future it might be possible to apply a repellent to bark to deter stripping. It will probably only be feasible to treat the lower trunk and the expense will rule out applying it to more than a few most favoured trees. There is, too, the prospect of spiking their guns another way. Contraception is a real possibility to limit numbers painlessly to protect vulnerable woods, and research on the grey squirrel 'Pill' is in progress. It may even help redress the widening imbalance between native reds in Britain, now down to an estimated 160 000 and continuing to decline, and greys already up to 2.5 million or so after just 120 years in residence.

Grey squirrels have had their chapter. They occupy more attention than most other management jobs, and perhaps have enjoyed undue attention here. When I visit the wood at intervals in the summer to check on the squirrel situation, other jobs are always added in. A frequent one is coping with the latest offering of rubbish tossed over the fence or dropped at the gate. What people throw away never fails to surprise, especially when they must have driven miles to get to our entrance when a municipal rubbish tip would be much nearer. At least the squirrels never try to cover up their deeds.

13

Over the fence

In the 1950s Vance Packard's 'Wastemakers' laid bare our throw-away society. The mountains of refuse produced day after day are the less than discreet hand maiden of our materialism. Interestingly, what seems to concern us most is where rubbish is disposed of rather than the enormous quantity itself. The dislike of litter, and the about-turn from litter bins at lay-bys to 'Take your litter home' signs, reveals the better strategy of emphasising responsibility for waste as well as saving the cost of collection. But rubbish is still thrown away in our countryside in huge amounts, despite municipal dumps, bottle banks, and regular refuse collection. And it seems that waterways and woodlands are prime sites. Even the unthinking reckless disposer who avoids taking rubbish to the dump or is simply too lazy to do so displays an ambivalence. He is anxious to see the very rubbish thrown away in the countryside at least hidden from view as it sinks to the river bottom or is tossed into the thickest undergrowth. Of course, sometimes this is deliberate to cover the traces of a theft or act of vandalism. The entrance to the wood has not escaped this late twentieth-century pestilence.

The most surprising rubbish has been the most recent. The week before I began to write this chapter, 20 fire extinguishers were thrown over the fence each side of the gate, a half-dozen to one side, the rest to the other. Ben and I gathered them up and hid them behind a tree as part of the continual attempt to keep the entrance tidy and deter such vandalism. The extinguishers were a mix of metal and plastic canisters, each with a short, black trumpet-shaped nozzle, were past their sell-by date and were emptied of carbon dioxide. The firm of suppliers in Wiltshire was easily identified but not the owner. Why throw away old fire extinguishers and why drive all the way to our wood to do so? I have no answer. At the same time, a cardboard tray of duck eggs also turned up. I have no explanation for these either, nor for the sudden departure of the same fire extinguishers six months later on New Year's Day 1995. Revellers had been at the entrance and it was probably they who had fixed a party-time red light bulb to each end of the wood's name board. These peered out, unlit, like some poor frog's bloodshot eyes. Whoever was there broke the crude sawing horse we used for cutting firewood, kicked around a couple of logs, and didn't tidy up their rubbish: only the extinguishers were gone.

The list of other debris thrown, discarded, abandoned, or dumped is long. Condoms and crisp packets vie for the most common. Tyres and sump oil are fortunately infrequent as are spent paint tins and other DIY detritus. Car ashtrays are turned out, sticky sweet papers and all, and girlie magazines thrown away. Bags of garden waste turn up from time to time, and so have two electric fires and one small fridge. As well as today's morality, pastimes, and living standards revealed by this catalogue, the electronic world has turned in one computer keyboard complete with umbilical cord, though not its monitor. All these have been inside the wood. So far we have been spared another pile of used tyres which Alistair told us blocked the entrance some months before we bought the wood. Also there has only been one theft as far as we know—not, of course, counting the brief departure of the grey water tank—and that was some nicely cut beech logs piled near the entrance ready for sawing and splitting for firewood in the winter. Whoever took them only pulled out the smallest leaving the larger logs behind in a tumbled down heap. At least they didn't move them to the other side of the wood like Pooh and Piglet did with Eeyore's efforts at house building when they found a dishevelled pile of sticks and logs. And, of course, it was a good turn that Pooh and Piglet were intent on doing their friend.

Litter and rubbish in the countryside spoil our picture of Merry Olde England. They spoil the Victorian idyll captured, indeed

perhaps epitomised, in the delightful illustrations embellishing Kate Greenaway's children's verse. Unpolluted land, unhurried work, carefree and timeless, with real labour for real reward as season follows season, contrast with our throw-away society, twentieth-century haste, and the perpetual stress of phones, faxes, and frustrations. It is easy to indulge fond sentiments of a past age and to hope it survives as we picture it in some corner of our green and pleasant land. Nearer the truth is that one scourge has replaced another. Two hundred years ago England's countryside, even Constable's countryside, was littered with paraphernalia to kill, maim, and generally debilitate the casual visitor on private land as well as those with malicious intent. Owners went to extraordinary lengths to keep people off their property, to prevent any kind of trespass. In large part this was to deny the poacher his quarry. Saw-toothed man-traps and firearms such as spring guns tied to branches, or propped up on sticks, replete with trip wire to trigger or other booby-trap mechanisms, defended the squire's estate and sport. Notices announcing 'Spring guns set here' were displayed as a deterrent. All too often the innocent and unsuspecting suffered, and throughout the eighteenth and early nineteenth centuries their death toll rose and rose. Such guns were not only set to catch poachers of game, but in graveyards even snatchers of corpses intent on selling the dead to the living for lessons in anatomy. The slaughter of the innocent—and in those days all but the propertied classes considered even poaching fair game and hardly a crime though the penalty was transportation or even execution—was very largely a rural blood spilling. Indignation at this appalling state of affairs surfaces in Cobbett's writing when he exclaimed, on seeing a farm sale include 'several steel traps and spring guns':

> And that is the life, is it, of an English farmer? How long will these people starve in the midst of plenty? How long will fire engines, steel traps, and spring guns be in such a state of things a protection of property?

The real blame for this wanton execution lay with the laws allowing only those in possession of large estates to kill game. It was even illegal to eat game, at home or anywhere else, unless it was a gift from the squire. By the 1820s these invidious laws were being attacked with special attention paid to ridding the countryside of the undiscriminating engines of execution. Lord Suffield led the fight for abolition pointing out, as many others had done, that a spring gun or man trap could show no mercy, could not listen to special pleading, could not know that the poacher had a family of ten to feed, or that the wanderer was simply lost, or a stranger to the locality. Twice bills came before Parliament only to be defeated, and it was not until 1827 that the spring gun finally became illegal.

Trespassers today may still be prosecuted but death no longer lies in wait for the wayward rambler. Indeed, today, damages can be claimed in the event of an accident, even by someone uninvited and trespassing on another's property, if inadequate care can be proved. From time to time even spring guns, quite illegally, are still set and reports of tragic accidents make the local press. Sadly, they are almost always suicides. No man-traps or spring guns are set in our wood, we just put up with the modern evil of litter and rubbish. The thoughtless dumping of rubbish saddens rather than threatens like yesterday's malice towards trespass. Both are defined by the boundary of the property and clearly marking this has from time immemorial established where ownership begins and infringement of law starts.

The four sides of our nearly square wood have four quite different boundaries. Top and bottom are the lane and railway respectively, and as immovable arbiters of location as could be found. To the south, next to Mike and Annie's field, is an old fence, a line of overgrown hazel with an occasional open-crowned ash and sycamore, and Mike's very new fence, mainly to keep rabbits off his organic vegetables. The least well defined boundary is on the north side which marches parallel for the whole of its 300 yards' length with adjoining woodland and scrub. It can be located at the top only by seeing where the good beech stops and the rough scrub and tangle of thorn and *Clematis* starts. In the middle part two or three old oak posts, getting on for 40 years old, prostrate or grotesquely leaning and with rusty wire still stapled home, mark the edge; while at the bottom the last planted row in Taid's wood forms the demarcation line.

Each of these boundaries requires different attention.

Where the wood fronts on to the lane an effective hedge has grown up. It has gaps and holes maintained by passage of roe deer, but is solid enough discouragement for people. The council cut the verges once or twice a year and we trim back the hazel and

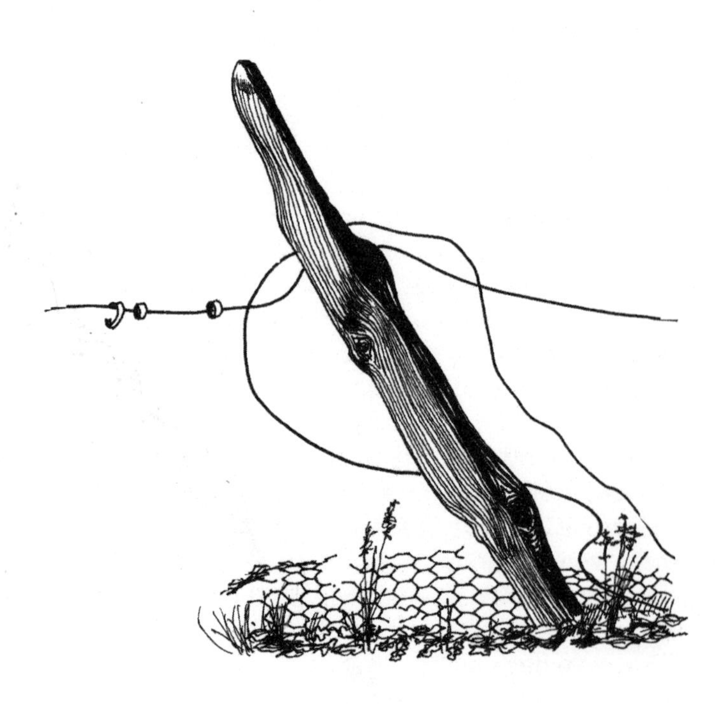

thorn that encroaches onto the lane. Our battery powered hedge trimmer holds enough charge to do about a quarter of the length at a time. Fortunately, overhanging branches are not serious and certainly no impediment yet to passing vehicles. If they were, both the Local Authority and the Secretary of State are invested with statutory powers to order their being lopped or cut back; similarly an encroaching hedge. We have been impressed by the Council's attention to the lane's condition. Each year workmen clear out the small ditches spaced at intervals to drain off water into the wood. There are, of course, no proper side drains and culverts for such a narrow, unimportant thoroughfare. Even so, twice in the time we have owned the wood the lane has been resurfaced. Each time care is taken to fill holes and uneven parts first and then some weeks later the new tar and chippings are laid.

The eastern boundary at the bottom of the wood is with British Rail or, as it is now, Railtrack. It accumulates its own rubbish of bottles and beer cans thrown from passing trains. Some of the fence by the railway is diamond-link but most is five strands of wire supported by concrete posts. It's great for kids to swing on and watch trains go by, if not entirely safe. British Rail have a duty to maintain it.

I have little to do along our boundary with Mike and Annie now that they have erected their own fence. This has been put about 3 feet inside their legal boundary. One day I would like to lay the hazel, sycamore, and occasional thorn to create a hedge to run along this side. This operation has never been done and would be for appearance's sake only. Mike doesn't plan to keep livestock, and a laid hedge is no barrier to rabbits or a determined deer. Facing south-south-west, this side of the wood catches the wind as it whistles unhindered past Bramdown house and over bare fields to deposit old fertiliser bags, plastic, and other bits and pieces. It's not serious, nor really a nuisance, just a noticeable feature of this most exposed boundary.

The northern boundary running contiguous with the scrub and woodland next door most needs maintaining in the sense of defining where it is. All too easily the line is blurred by new growth and, even now, it is only apparent if one knows what to look for. Each year a little trimming and cutting back helps to keep open some kind of sight line. There is, though, no clear sign or change to show when one leaves the wood and crosses into our neighbour's. This lack of landmarks serves to remind of what the

Bible inveighs against at least half-a-dozen times as a particular evil, on a par with those who cheat customers with fraudulent scales or doctored weights and measures: the moving of boundary stones. Twice in the second book of the law, Deuteronomy, and twice in Proverbs are warnings against moving a neighbour's boundary stone or one which is ancient. Job, replying to his 'comforters', lists this deed in the same breath as sheep stealers, and those who prey on widows and the poor, when he asks why God does not judge such wrong (Job 24.2). We have no marker stones, only a map showing a line, and some old, rotting oak posts indicating where this boundary is supposed to be: so I hope we can avoid this transgression.

It's not necessary to walk the boundaries regularly, though most parts get visited once a year. This is fortunate as it is much overgrown next to the railway, and because the extent of stinging nettles has much increased owing to thinning out the pines and the felling and replanting at the bottom. On many visits, even with thick trousers, tingling knees remind one of the nettle's needle-sharp ability to penetrate and fire off its stinging cocktail of histamine and acetylcholine. The sensation remains for the rest of the day. Not a large rash raising white blisters of urticaria in need of a dock leaf, just a tingle calling attention to what one has been doing.

Maintenance of internal divisions in the wood is confined to an annual swiping of grass and nettles along tracks and rides. Mike has kindly done this for the last few years, initially in exchange for the treeshelters that were surplus after finishing the planting of Taid's wood. From time to time the beech beside the main track and turning bay are specially high-pruned to maintain access and to reduce shade and encourage flow of air to help keep the track surface dry. Neither British Rail, the keeper, nor I want to get stuck in the mud. Ideally, tracks and rides in forests should be wide with bays scalloped every so often to keep them open and also encourage wildlife. In a small wood this extinguishes a great deal of ground and we have confined this conservation work to the central intersection, which has been opened up, and by deliberately keeping back from the edge of the track the final row of planting in Taid's wood.

The wood has one access from a public right-of-way, the bellmouth entrance from the lane at the top. Two five-bar gates, one large and one small, swing on large oak posts. They are old, weathered to a greyish-olive or even greenish where blue-green algae have found the damp wood congenial, and a little moss has taken hold in moister crevices. Ivy is beginning to clamber up the oak posts. The gates remain serviceable, they swing open freely, and meet at the centre of the track where they fasten together with a rusting iron latch. The years have worn the hinges, and the bigger of the two gates sags a little and no longer aligns quite with its smaller counterpart. A slight lift, like a friendly arm supporting an ageing relative, is needed to help it home. A heavy chain, sporting three hefty locks—British Rail's, the gamekeeper's, and ours—secures the gates. To the side of each gate is a simple stile of two bars for climbing over. The gates, positioned 25 feet into the wood from the lane, provide an opening of about 14 feet. At the lane itself the bell-mouth is some 50 feet wide and the whole metalled entrance forecourt, gate arrangement, and short track and turning bay inside the wood are sufficient for a large timber lorry to manoeuvre—at least it is supposed to be. Once or twice vehicles have backed into the gate, bending the iron latch and into the side fence and cracking some of the bars, but not causing damage beyond repair. The gates were put up by the Forestry Commission and have lasted well, although the adjoining fence of treated

softwood has finally rotted. Repairs were done with Douglas fir poles cut from the wood and some Lawson cypress from our garden. This cypress, which is so common in suburbia as a garden hedge, produces an extremely durable timber, only equalled by sweet chestnut, and will last untreated as a stake in the ground for a quarter of a century or more. The Douglas fir is only moderately durable and will probably last about ten years.

Stephen helped with the repairs to the rotting fence on a frosty, crisp, and brilliantly clear day in late December. Earlier in the year the cypress poles had been stacked at the bottom of the wood and the first job was to select some of them for making into stakes or using as rails. These were manhandled up the 250 yards of gently sloping track to the entrance. The heavy damp poles had us both breathing hard, which the cold air condensed into misty puffs lit up by shafts of sunlight slanting low through the trees that winter's morning. It was still, and each puff lingered for that moment longer before evaporating into the mists of time. We lingered too, pausing every so often to catch our breath before the next effort and another 50 yards brought us closer to the gate. The Douglas fir poles were cut near the entrance from standing, but dying or dead trees suppressed by the success of their neighbours. With the woodwork assembled, the next task was to point the posts, which we did laboriously with hand axes. This skill, long dormant since acquiring it in scouting days, raised blisters despite the leather of Ruth's gloves. The continued exertion continued the puffing and we kept warm. A crowbar opened holes for the stakes, these were driven in as far as they would go, and the horizontal rails nailed in place. I attempted to chamfer the ends to splice them neatly together, but the angles kept coming out wrong. It was invigorating to work on such a glorious morning. At the end of it even the newly fashioned fence which was taking shape looked acceptable in the bright sunshine. We returned on New Year's Eve to finish the job. The brilliance and frost were gone, replaced by the penetrating cold of chilling dampness a degree or two above freezing. We hurried to get finished. The repaired fence looked good and has stood for three years so far.

The threshold sign at the entrance we inherited from the Forestry Commission has also lasted quite well. It had to be re-erected after it blew down in a gale in December 1989 which revealed that rot had decayed the base of the posts. The letters of the wood's name and the warning about no public access because of reserved sporting rights have been repainted, but the sign board and posts need doing. The flaking paint of the Commission's sombre sea-green is unfair both to the former owner and the sign's presentability. Recently it's had its long awaited coat of paint and change of colour.

Regular maintenance of the gates is confined to oiling the locks and the hinges, though the top ones cause a problem. Like many forest gates, these hinges have been deliberately fixed upside down to prevent the gate from being lifted off. It also prevents easy lubrication with oil since we don't have one of those old cans that look like Aladdin's lamp, which squirts the lubricant, not genies, by a thumb-operated pump. Along with the grease gun, it was once part of every driver's tool kit. But we still have the gate and in our ten years of ownership no one has tried to gain vehicle access without permission. One can never be around all the time to see who has tried, though there is no doubt that poaching and theft, ageless country crimes, occur in the locality.

Two rights or easements are granted through the wood. One is of long standing to allow rail staff to visit the grey transformer and carry out track inspections. They use the main track and have their own padlock on the gate. The other is recent and was granted to Southern Electric to lay a cable across our land from Mike and Annie's to the transformer. After a site meeting we agreed that it should run underground and follow the line of the track along the bottom of the wood. We have been paid 0.34p per square metre for this permission, the standard rate agreed with the National Farmers Union. Underground cables sterilise a 2-metre (6 ft.)-wide strip. Ones erected overhead on poles, although usually cheaper to install, affect a much wider swathe to avoid interference from trees which, of course, is what we are trying to grow.

As small landowners we acquired more than a lot of rabbit burrows, a lot of trees, and an awful lot of time. We acquired all the duties and responsibilities as managers of a tiny country estate. We also acquired other people's rubbish. But there was something else. Owning even our small wood opened up many opportunities to share it. And that has been worth all the rubbish and the chores of attending to the little estate's duties. First, however, one significant event recently took place which eventually led to extinguishing altogether the ill-defined northern boundary.

14

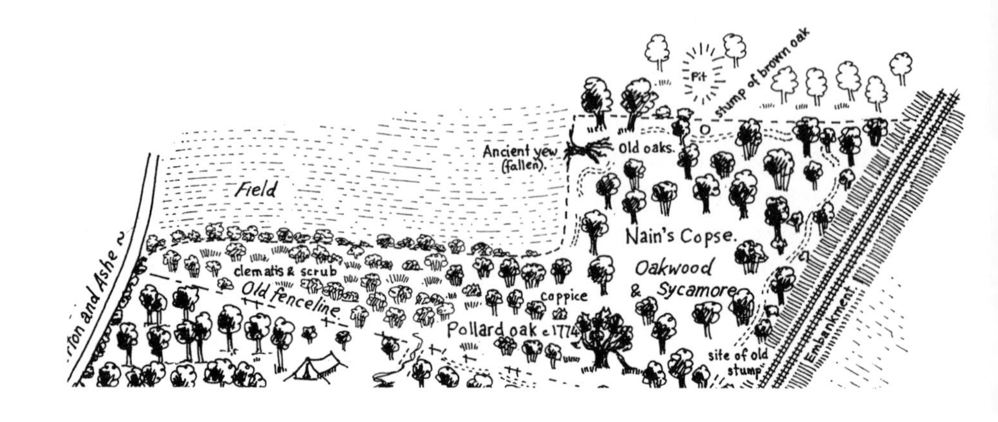

Nain's copse

Adjoining the whole length of the wood's northern boundary is another, smaller woodland of about ten acres. It fills a triangle of land with two completely straight sides, the ones with our boundary and with British Rail's, which converge almost at right angles at the point of BR's grey transformer. The third side traces a distorted 'W' like the stars of Cassiopiea. The wood belonged to the Steventon Estate and was never leased to the Forestry Commission.

In the ten years we have been neighbours no work has been done in this wood apart from coppicing a tiny panel of hazel, perhaps amounting to a dozen clumps. It has been used for sporting and occasionally a temporary deer seat would appear a few yards inside their boundary to overlook the young trees of Taid's wood, where deer would browse and sight lines were easy while the trees were small. In the 1987 storm a huge birch blew down across the track that is a continuation of ours and which runs along the bottom of this adjoining wood beside the railway. The tree still lies where it fell, as does an oak at the far end near Waltham lane. The track between the fallen trees has become much overgrown.

Apart from occasionally pausing to admire a splendid pollarded oak that guards their boundary (the tree spreads in all directions and can boast a 14-foot girth), visits to our neighbour's wood have been confined to investigating the threat grey squirrels presented. The old oaks, several full of cavities and crevices, and numerous sycamores make ideal squirrel habitat and the whole wood serves as a grey squirrel holding area. Rabbits, too, abound, but for us, though not for our neighbours, they are less of a nuisance, and have left quite untouched many fine clumps of primroses. One other interest had been aroused. When Rod Stern and his fellow bryologists assayed our wood for its holding of mosses, lichens, and liverworts, they strayed into this adjacent part and it was there that they discovered the first recorded occurrence in Hampshire of a tiny moss called *Platygyrium repens* growing as an epiphyte. It was on some leaning ash and field maple about ten yards beyond our boundary. On a later visit Rod showed me their discovery, its distinguishing feature and hence rarity discernible only under a ×10 hand lens. This small neglected wood held considerable interest for us, harbouring not only rabbits and grey squirrels but also some great oaks and a tiny rare moss.

In 1990, agents for the owner of this wood, which had recently changed hands as, in consequence, had our lessor, wrote to say that it was being offered for sale along with the large field on the south side of the wood. Once or twice before, we had enquired about converting our leasehold into freehold, and they knew of our interest. The adjoining wood and field were being offered in either one or two lots. John was reluctant to invest much more in forestry and Margaret and I didn't wish to proceed independently. We did, nevertheless, put in a low offer, more what we could afford than probably what it was worth, to express interest. Nothing came of it. It was at this time, however, that Mike and Annie were able to buy about one-third of the big field and become our immediate neighbours to develop their organic vegetable holding. Melvyn bought the middle bit of the field next to theirs, and has plans to cultivate thatching straw. For us the adjoining wood, full of pests, possibilities, and peculiarities, continued to languish along with our reversion to freehold.

More intensive keepering in our wood, in particular the deployment of the 44-gallon drums full of winter feed for pheasants, led to renewed contact with the lessor's agents. As I complained in Chapter 12, feed good for pheasants is also good for squirrels, and the arrival of the drums precipitated in the spring following the worst damage we had experienced. We could not insist that this feeding stopped, since our lease specifically excludes any control over the sporting interest. The keeper was as helpful as he could be, delaying the start of winter feeding until mid-November and promptly removing the drums in early February when the season closed. Correspondence with the agents seeking to resolve the matter eventually led to renewing the proposal that we buy the adjacent wood along with our freehold including, obviously, the sporting rights. This would solve the problem of control over the sporting and dispose of what was a tiny outlier of woodland for the estate. Moreover for us, the price being asked was less than it had been in 1990.

We met with the owner to agree terms in March 1994. John and Gill were only interested in acquiring the freehold of the existing wood leaving to Margaret and me the purchase of this neglected wood on our northern boundary. A hiccup in negotiations meant that we did not acquire quite all of the new area, the owner deciding to exclude the far corner wedge of three acres next to the railway and Waltham lane. The agreed price was modified accordingly. My mother, Nain to our children, who had long since fallen in love with our wood, helped with the purchase. Part of the new wood is of greater pedigree than our existing one and is clearly shown on the 1876 Ordnance Survey maps, but it has no name, not even on the largest scale maps we could find. No name appears for it on any subsequent map either, nor on any of the deeds, transfer, and legal documents accompanying the purchase. So we have named the wood 'Nain's copse' in gratitude, and to complement Taid's wood which it adjoins. The Welsh is also much less of a mouthful than calling them 'grandmother's' and 'grandfather's' copse and wood respectively, and anyway, the boys have always known their grandparents on my side as Nain and Taid. Nain's copse legally became ours in early August.

In agreeing to buy Nain's copse I had already looked carefully at what we would be getting. The price per acre was quite a bit more than what we had had to pay in 1985, even taking into account that we were buying the freehold, worth itself about £200 per acre, and not just the leasehold. The woodland was very different from what we already owned and was itself of two very different parts.

The smaller of these two parts forms a spear-shaped triangle of ground of about 2.5 acres, and consists of scrub. This denigratory term has been much over-used in forestry and all too often has labelled what is perfectly satisfactory woodland regrowth in the thicket stage simply awaiting cleaning. True scrub, from a productive forestry point of view, is where there are either no or very few woody plants of tree potential. This smaller part of our new woodland was genuinely scrub: it contains no significant trees, just the inevitable *Clematis*, several hawthorns, some hazel clumps, one or two straggly birch, some field maples, a whitebeam, and a few small hedgerow oaks. The *Clematis* drapes over the hedge near the lane and assaults the planted beech nearby. Added to this, and probably explaining the lack of tree growth, are numerous rabbit burrows. Its one redeeming feature is a glorious display of primroses right in the middle. Such scrub appears to be the result of neglect. Early maps do not show this land as wooded, just a blank triangle, and it seems clear that no tree planting has ever taken place; at some point the land was simply abandoned and what we have bought is the result. It is a good example that abandonment alone does not always ensure development of woodland, at least not for a great many decades.

The larger part of Nain's copse, some 4.5 acres, is a coppice with standards. The coppice is an unconventional mixture of sycamore and a few ash with a little hazel on the fringes. It can no longer truly be called underwood since it was last cut some 40 years ago or more and has grown up into tall stems, frequently matching the height of the oak standards. This stored coppice, as it is called, is now a dense crop and offers the distinct possibility of thinning and conversion to a fine high forest stand of these two excellent white woods. It could be re-coppiced, but the neglect has been rather too long, and many of the stems are already halfway to sawlog size. Some of the sycamore received the attentions of grey squirrels in the same year, 1993, when our wood suffered such widespread damage. Bark stripping was confined to the base of perhaps two dozen stems. Scattered through this stored coppice are 97 oak standards varying in diameter from a small 35 cm (14 in.) to as much as 129 cm (51 in.), although most range between 45 and 80 cm. It is these oak standards, again probably not touched since the war, which caught my attention. They explain why one Saturday in late February, a few weeks before we were due to negotiate the purchase, Ben, our youngest son, his friend David, and I went to the wood to count and measure these veterans. The tally as indicated amounted to quite a large number and just over half were of the marketable size of 55 cm in diameter or more, equivalent to 21 true inches or, more importantly for hardwoods, 17 inches quarter-girth. It was quite an impressive collection of oaks, although a few had signs of rot at the base and a couple had had their tops blown out, and they were desperately in need of thinning out just like the rest of Nain's copse. What impressed Ben, from his later recollections of the day, were cold feet and frozen hands as he stood around holding the girthing tape for Dad as each oak was carefully measured: there is nothing quite like hanging around on a dank February morning, a degree or two above freezing, for chilling to the bone. The large number of oak standards and the dense stored coppice of ash and sycamore were persuasive arguments for buying this wood, but I was anxious for a second opinion.

I turned to John Newcomb, through whom we had bought the original wood nine years before, and he kindly came over so that we could walk together through the new area. On the whole I was reassured, particularly since the purchase included acquiring the freehold over all our land and extinguishing the long northern boundary with this adjacent wood. We decided to proceed. A meeting was arranged with the owner to discuss terms. John, my brother-in-law, who would be sharing in obtaining the freehold, also came. Thus we became owners of 30 acres of woodland in all; and this time for ever, not just the next 967 years! Though, as nice as the wood is, there is an even better place where one really can spend eternity.

One matter of estate management needed attention here and now. At the time of negotiating the purchase, we agreed on the map the boundary between Nain's copse and the 3-acre triangle retained at the northern end as a line which continued the straight direction of the field edge to the point where it intersected British Rail's fence. This was surveyed and stakes with white-painted tops hammered in at about 30-foot intervals as temporary markers. We found this line just excluded from our ownership an old bomb crater, a splash of wild snowdrops, a fine field maple, and a magnificent pollard every bit the equal in age and stature to the one in the top corner of Nain's copse. There is no point in anyone cutting it, so although it's not ours there is every reason to hope it will remain. Our responsibility is to manage well the woodlands we own, over which, really, we are no more than stewards caring as best we can for the present years of their existence.

Nain's copse added not only more timber and increased size, but in the spring small patches of bluebells, primroses, a couple of clumps of cowslips, some unusual false oxlips, and another sea of Dog's mercury. Doubtless further visits will add to the inventory of flowers but, as this chapter indicates, we were pleased to acquire this new wood. In many ways it has completed and balanced our small forestry estate. In the 30 acres there is some scrub awaiting attention, a young plantation (Taid's wood), a maturing crop in mid-life (the pole-stage beech) with many decades' growth still in prospect, patches of Douglas fir not far off final felling, some mixed broadleaves almost as problematic as the new scrub, and now a dense coppice with oak standards rapidly passing maturity and in urgent need of attention. What exactly that should be has not been easy to decide, so Nain's copse posed a silvicultural teaser too.

The recent neglect of Nain's copse is easy to see and our objective of wanting to encourage wildflowers, and wildlife in general, along with some yield from the timber is easy to state, but not so easy to achieve. Three main options appeared possible. The whole copse could be felled and regenerated, since most of the oaks were close to or past maturity, and the stored coppice of ash and sycamore are just the right size to yield plenty of pulpwood. The mill at Sudbrook in Gloucestershire recently expanded capacity and a good market exists for hardwood pulp. A clearfelling would maximise an immediate return and fully open up the long-shaded forest floor. It would, however, totally change the wood, and the view from the railway, and it would consign another area to replanting: there would be little advantage in resuming coppicing since sycamore predominates, with perhaps less than usual to gain for wildlife or as a forest crop. The second possibility would be to begin felling and regeneration either among several scattered groups of trees or in two or three stages over the whole wood. This would maintain better forest cover and add diversity to the wood's structure, but inevitably untreated parts would persist in a state of neglect. The third option would be to thin out the overstocked oak standards and recruit the best of the stored ash and sycamore stems to high forest. Although substantial openings would result from where the large spreading oaks were cut, the copse would continue to look like woodland, it would help the ash and sycamore become fine trees, it would bring needed light to the forest floor, and delay regeneration for two or three decades.

By this time the adjacent planting in Taid's wood would be about 40 years old and so make any substantial work in Nain's copse far less obtrusive. Even after such thinning there would still be more than 50 oak standards which achieves the right density of about 10 per acre or 25 per hectare. I decided for this option, with the suggestion from John Newcomb that the oak standards were thinned out and marketed first, followed, as a separate operation, by thinning of the stored ash and sycamore.

The girthing of the oaks that was done in February now came into its own and 44 of the 97 standards were selected and marked

for felling. These were mostly the bigger older ones over 55 cm in diameter, but the few real veterans, including the great pollard guarding the top corner, were left and will ever be so for posterity and for their tremendous wildlife conservation value. Each standard to be cut was painted with a white spot on its south side. It will be a tricky job felling these large heavily branched trees and avoid damage to nearby ash and sycamore. Long before worrying about this stage, one had to apply for a felling licence. On the licence application the name 'Nain's copse' was used for the first time and, in due course, it might even appear on the larger scale Ordnance Survey maps—a little history in the making!

Marketing large oaks is quite different from run-of-the-mill pines or similar bulk volume timber crops. Each log will have a particular value depending on its diameter, length, and straightness, and whether it has any defects, most notorious of which are decay in the heartwood and a problem called 'shake'. Shake is the name given to long cracks in the timber that either fan out star-shaped from the central pith or follow an annual ring. They can run for yards up a log and render it nearly worthless as the log falls to pieces when sawn. This defect tends to be associated with gravelly and sandy soils, but no stand of oak is ever entirely shake-free. The trouble is, like many cancers, shake cracks cannot be detected from the outside and foresters don't have log scanners, at least not yet. Occasionally blemishes can be turned to advantage: oaks with large burrs—burr or pippy oak—are sometimes specially sought after for cabinet work, and ones with wood stained brown from infection by the beefsteak fungus, *Fistulina hepatica*, can add pounds to the value. For all these reasons mature oak is rarely sold standing, in the way the pines were, but felled and debranched, or 'dressed' as it is called, and left on the ground for potential buyers to inspect. This is selling 'at stump'. Each log is numbered and the buyer can see clearly what he is getting. Provided there are not too many defects and blemishes, this approach will reward the owner with a much better price compared with the riskier business for the buyer of purchasing oak still standing. Although felling the trees commits them to being sold, it is not quite the buyer's market it might seem. Good oak are always in strong demand and logs can be safely left on the ground for many months without deteriorating. Indeed they will be slowly seasoning, which is just what will be done with them when taken to the mill, as unwieldy stacks of oak logs at the country's sawmills testify. There is also one other marketing difference compared with the pines.

In 1990 we sold the pine and estimated the total volume of wood in the 2000 or so trees to be 636 cubic metres. Oak is still sold in the traditional measure of 'hoppus feet' which I described in Chapter 10. The buyer will not be interested in 61 cm diameter, but its equivalent of 19 inches quarter girth and a log volume of say 65 hoppus. Bargaining over the price is done at so much per hoppus foot. In the case of oak this varies from 50 pence for poor material suitable for firewood, to about £1 for fencing quality, £3–4 for good sawlogs called 'planking quality', to as much as £10 for high grade, wholly blemish-free veneer butts. Only walnut and one or two specialist species ever make these higher prices. Large sound oak are valuable: one first-class tree can be worth several hundred pounds.

Once again I turned to John Newcomb for his experience to arrange the felling and sale of the oak. It is not a job for the uninitiated and my long research experience of how best to grow and care for oak was no qualification for how best to market them. Like this book, it's one thing to write it, quite another to market it successfully.

15

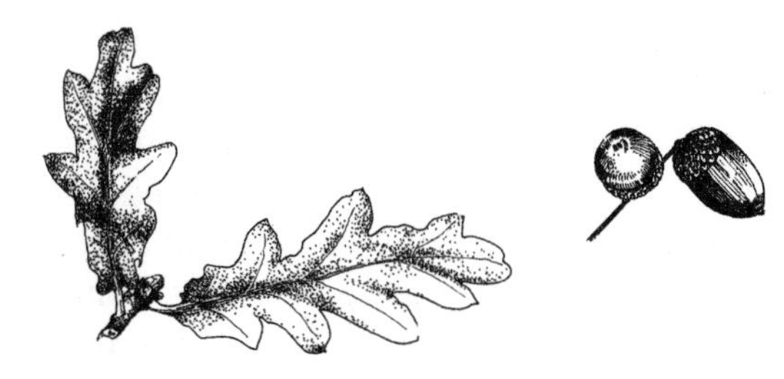

Oak

The oak standards in Nain's copse were ready for thinning. The dense canopy of the copse with tall, drawn-up sycamores jostling with the spreading crowns of the over-stocked oaks heavily shaded the woodland floor. A few emaciated primroses struggled to glean what little sunlight penetrated, their leaves distended, pale and flaccid, and even the aggressive Dog's mercury hadn't colonised everywhere. Parts of the floor were bare with dead leaves, twigs, and fallen branches. Clumps of ash and sycamore coppice, which hadn't seen open sky either, languished etiolated[1] in terminal decline, denied their share of radiation and warmth. There was little point in delaying thinning, and a start was made shortly after Nain's copse became ours.

Forty-four oaks were selected for felling and painted with a white spot. On the day after doing this I was smitten with indecision about two of them. One in particular possessed a large, immensely spreading crown with only a short butt, albeit of good quality. It looked more like a field-grown tree than one nurtured in forest conditions. I had marked it for thinning, but left two adjacent, thinner, but taller and straighter oaks. Further reflection changed my mind: the big tree should not be cut. Cutting this massive oak would create a huge hole in the canopy and cause much damage as it fell. Back in the wood next day and armed

[1] Pale and elongated due to lack of light.

with a hand-axe the glistening white-painted spot was chipped off bit by bit, the spot rubbed with mossy soil as further disguise, and the two adjacent oaks branded with paint instead. While in Nain's copse, a veteran oak next to the great pollard in the top corner also received a reprieve and more paint was scraped off and more dirt rubbed to obscure the mark. The pollard, this veteran, and a third oak, which together were probably old field boundary trees, can continue as the triumvirate they have been for more than 150 years. Neither they, nor the great spreading oak, which John White estimates as being 220 years old, will be touched in my lifetime.

After the change of mind I tried out the amended selection on Gary Kerr, silvicultural expert and the senior author of the Forestry Commission's book *Growing broadleaves for timber*. He was at the wood a few days after the marking was done, along with heads of research branches from Alice Holt Lodge whom Margaret and I had invited for a late summer picnic. Gary concurred with leaving the big spreading oak though he would not have taken both the adjacent smaller ones. We both liked the three old field boundary veterans. I didn't return a second time to chip and paint. I should have done.

During the felling operations an oak on the edge of the old boundary with Nain's copse, just opposite the great pollard, was also cut. It bore a white spot, hence the reason for the mistake, but which had been painted in 1986 to distinguish it from the trees retained by the lessor on the land which became Taid's wood. Eight years of weathering had not been sufficient to avert the keen-eyed attention of the fellers! Interestingly, this tree was of good form, but possessed some features of Turkey oak, such us deeply scalloped leaves and no epicormic branches on the stem, and would therefore be of less interest to buyers. On discovering this mistaken felling, and since the work was still going on, another tree was given a reprieve, once again chipping off the paint mark, to leave standing as a substitute. It was not strictly necessary to do this since one can fell up to 2 cubic metres of timber per quarter without permission. The smaller oaks being felled were about this volume.

The felling licence granted permission to thin 43 oaks and one ash of an estimated total volume of 110 cubic metres. John Newcomb arranged the cutting with Tony Whatton of Hartwell, Northants, agreeing a price per cubic foot (hoppus foot really) on felled measure: that is based on the volume of oak logs cut and trimmed ready for sale. As with all marketing of quality hardwood the eighteenth-century system of measurement was used. Tony agreed to do the cutting at a rate per hoppus foot of log produced which averaged out at £14 per tree. We also agreed with him that the lop and top from the oaks, with their broad crowns and plenty of branchwood, should be cut and stacked in cords—piles of 4-foot lengths, 4 feet high, and 8 feet long. This amounts to 128 cubic feet, or roughly 2 tons when felled, and was originally measured by a rope of fixed length, hence the 'cord'. The cordwood would be sold separately from the main logs.

Thinning began in late September and took about a week and a half. The weather was perfect. Sunny days and light winds of that autumn will be remembered for the glorious display of reds, yellows, and browns as leaves gently aged (senesced) on the tree: for once not torn off, but able to depart gracefully. No harvesting machine cut the oaks, just highly skilled tree fellers equipped with large chainsaws. In their different way their work was as impressive as the Hymac had been among the pines two and a half years before. The oaks were cut only an inch or two above the ground, their direction of fall controlled by judging the balance of the tree's crown, and by the skill of placing sawcuts and inserting wedges as cutting proceeded. The feller assays the crown, noting the heavy limbs, the lean of the tree, and other distortions displacing its centre of gravity. He has to cut with nature, not against, and can only coax the wooden giant to fall where it should. Judicious sawing and long experience are the only aids; no rope, no winch, and no machine to rescue from miscalculation or misjudgement. John Newcomb and I watched one such unwieldy fieldside oak being felled and land within inches of the intended spot. This skill is critical to getting the best out of such trees since poor choice of direction can not only damage remaining trees but also ruin the log itself. If a falling tree impales on a large branch or some obstacle, the stem can easily split in half, and the value be reduced to a tenth! Once on the ground more skill is required to cut off side limbs, called 'dressing the log', and to decide the point of crown break, and so where to cut the top off. Sometimes deeply forked trees have small logs converted from the giant limbs, though their quality is often indifferent. Rather like the diamond cutter, it is the skill of the tree fellers to cut, convert, and shape the oak trees to

fashion them into the best logs they will yield: they can make or break the operation. Like diamonds, too, the larger, finer, and more unblemished the end result, the more precious and the more valuable.

The thinned out oaks to offer for sale at stump, revealed that few had any signs of shake and only one even a trace of decay. Thus they could be marketed as a parcel of quality timber. The oaks were not all the same age. Most of them were about 115 years old and thus had almost exactly served their time by silvicultural lore which teaches that oak should be grown for 40 years for height, 40 years for girth, and 40 years for maturity. A few large oaks beside the field were older than 155 years, and may date back from the time of field enclosures, though they are almost certainly on a much older boundary. On this same field margin a long fallen tree, not closely inspected by me and dismissed as of no interest, turned out to be a very old yew, rotten at heart, but undoubtedly many hundreds of years old. It probably blew down in the '87 storm. More certain is the fact that, unlike the rest of the wood, and probably Nain's copse too, Jane Austen would have seen it. It may even have been a favourite. On a frosty winter's morning its dark green mass and red berries would herald another season, as evergreens do, for the traveller hurrying for warmth along Waltham lane. Pausing to catch breath on the long hill, Jane would see the old yew across the field in the far hedge as she turned in its direction to catch what little heat the tired rays of the low sun could offer. This not entirely fanciful linking of the novelist with Nain's copse, more sure if the fallen yew could tell its story, would please my mother. She read English at Girton, Cambridge, and has great affection for Jane Austen. As a boy I recall her sitting by the fire and re-reading again and again her favourite, *Pride and prejudice* which Jane wrote while still at Steventon. In later years she has come to prefer the mature reflections in *Sense and sensibility* from Jane's own later years in Chawton. The link pleases us, too. Many a delightful lunch has been enjoyed in Cassandra's Cup, the cafe in the old post office opposite Jane's Chawton home. Indeed, it was there that Margaret and I discussed buying Nain's copse, and where we lunched before picking up John Newcomb to go to the wood to agree the thinning of the oaks, and how best to market them.

Looking around Nain's copse suggested that only the yew had blown down in the great storms. No oak had succumbed, although a couple had lost branches and one tree had had one side of its crown ripped out. A great broken limb still hung forlornly, brown and dead, unwilling finally to lose its grasp. The stability of these oaks suggested sound rooting free of decay and penetration of soil to reasonable depth: they may even have been planted with the care Giles Winterbourne and poor Marty took with the pines in Hardy's *The woodlanders*.

> Winterbourne's fingers were endowed with a gentle conjurer's touch in spreading the roots of each little tree, resulting in a sort of caress under which the delicate fibres all laid themselves out in their proper direction for growth. He put most of their roots towards the south-west; for, he said, in forty years' time, when some great gale is blowing from that quarter, the trees require the strongest holdfast on that side to stand against it and not fall.

It is hard to prove whether such care has a benefit, though it is unlikely to do any harm. All too often damaging gales strike from an unusual direction such as the southerly quarter from which the great storm of '87 blew. It is clear, though, that one reason for wind blowing down trees in the more exposed parts of Britain's upland forests is that roots have aligned parallel to the furrows and ridges created when the ground was ploughed before planting. Few roots manage to cross such an obstacle, so the trees' support is one-sided and uprooting more likely.

Less conjectural were two more discoveries. The one ash felled was sound and white in wood, a good augur for the trees we planted in Taid's wood, since many fine ash disappoint on felling with dark stained heartwood. Such ash are unfit for any quality uses. The other discovery, to our great surprise, was a genuine 'brown' oak caused, as mentioned in the previous chapter, by infection from the beef-steak fungus. It was the first I had seen and John Newcomb highlighted it in the sale particulars. We would not have known about either the good ash or the brown oak without the trees being felled in advance of sale. As William Pontey wryly observed in his *Forest Pruner* 180 years ago 'An old oak is like a merchant; you never know his real worth till he be dead.'

The bottom one foot of the brown oak had decay and the fellers cut it off, but above the wood was stained a rich, nutty, chocolate brown across its surface. The dark colouring became typically streaky at the edges near where heartwood turned to sapwood. John advised cutting it through about ten feet along to see if the brown colour persisted to the first branch stub. The feller obliged, but the two ends did not separate and remained too tightly together for inspection: the sawdust produced did seem similarly brown. This brown oak was a bonus and would interest wood carvers, cabinet makers, and other specialised users always keen to get hold of this rare and attractive feature.

After felling was finished Tony Whatton numbered each log and measured the volume to prepare his invoice—I had agreed to pay him by the cubic foot. When the figures arrived John Newcomb and I went to check the measurements. A large shocking pink number proudly emblazoned the butt end face of each log. We selected about ten logs for checking including the brown oak which now boasted the number 20. Log length in feet was run out with a tape pinned at the butt and the mid point determined. Here the girth was taken. A quarter-girth tape was run round the log to measure its diameter in quarter-girth inches, just in the way prescribed by Edwin Hoppus over 200 years ago. Getting the tape under a heavy oak log, or any log for that matter, is difficult and at this point John produced a tool I'd not seen before in my 25 years of professional forestry. It was about 2 feet long—I'd better stay with imperial measure—curved a bit like a scimitar and with a pointed tip. Just below the tip was a deep recess cut in the thin narrow blade making it look like a large fish hook. This measuring or timber sword, as it is called, is inserted under the log, and it goes very easily, to emerge far enough on the other side to expose the hook. The girthing tape is dangled like a fishing line over the far side of the log and the metal ring at the end caught by the hook. The tape is then readily drawn under the log like a needle and thread to complete the encircling and allow the girth, or more usually diameter, to be read off. This simple device saves hours of scrabbling for the end of the tape, particularly difficult when confronted with large logs greater than one's armspan. We were well satisfied with Tony's measurements and his calculation of log volume using length and mid-quarter-girth from Hoppus tables. I paid his invoice and we included the data—feller's measure—with the invitation to tender for the parcel.

Invitations to tender were sent to six prospective buyers, mainly sawmillers or their agents across southern England. Being 'at stump' the logs were open to inspection and their measured volume susceptible to accurate checking. I've no idea whether potential buyers came swashbuckling with timber swords too! We advertised the logs separately from the cordwood since quite different merchants would be interested. The particulars stipulated that great care should be exercised when extracting the logs to avoid damage to the remaining oaks and sycamores, but otherwise standard terms and conditions were included.

One other preparation was high pruning and cutting back the main track down to the railway and along the bottom of Taid's wood to Nain's copse. There the track peters out by the transformer, it's as far as the British Rail staff ever go, and the long fallen birch blocked further progress. Ben and I spent a morning with chainsaw and axe cutting our way into Nain's copse through seven years of accumulated growth where the track once was. The birch was rotten and easily dealt with, but the vigorous hazel and inevitable *Clematis* took much clearing, tugging, and hacking back. Two coffee breaks later and by the end of a long morning access was restored and the old track partially reinstated.

In early November the tenders came in and all six potential buyers had bid. Just like the pines, they ranged more than 50 per cent from the lowest to the highest offer. The two best were within a few pounds of each other, and I was pleased that English Woodlands Timber Ltd were successful. I have twice been invited to talk to their staff about the work of Tear Fund and Tree Aid about how tree planting can help ease deprivation in the so-called Third World. For so many the necessities are firewood, or sticks and poles for building homes or fencing for livestock, or planting to prevent soil eroding or growing fruit trees; certainly not growing trees as a timber crop. English Woodlands have been one of the staunchest supporters of Tree Aid and have helped fund tree planting over a whole catchment at Matere, in Benin in West Africa. The price they paid for our oak confirmed it as a small, reasonable quality parcel, though not special apart from the interesting brown oak: the figure came to more than four times what we had got per cubic metre for the pine. The calculator revealed though that they had priced the parcel by the hoppus foot in true hardwood merchanting fashion.

One of the unsuccessful buyers chased up the brown oak butt hoping to buy it from English Woodlands. I think he planned to use it for restoring antique furniture. The oak tree at 155 years was an antique too, but the brown staining from the beefsteak fungus gave it the age to match, hence its value.

A tithe (one tenth of the income), perhaps the value of the brown oak, was set aside for use in the Third World. It has gone to support a friend, Rachel, working among Kurdish refugees in northern Iraq. In its way it continues English Woodlands' own concern for needy peoples, and the nearest foresters can get to the farmers' 'Send a Cow' programme! 'Send a log' would not do much good, but sending something to the desperately needy refugee peoples of the world, of which the Kurds represent just a few of the estimated 46 million in the world, helps in a small way. It is a privilege for us to share our many blessings, and even those from the wood, though they are but crumbs from the wealth we all have in the West.

English Woodlands employed Tony Whatton to extract their logs. Again a first-rate job was done with hardly any of the remaining ash, oak, or sycamore trees barked at the base as the heavy cumbersome oak logs were winched in and then pulled out by tractor. Little rutting occurred, despite the very wet week in early December, in dragging the logs all the way from Nain's copse to the entrance gate. The skimmed mud of the track resembled blancmange. A wary roe deer or two tested it, their slots etched as if in fresh snow. At the gate the logs were not piled high like the pine had been, but laid either side of the track and turning bay with each vying for attention. The brown oak had pride of place at the head of the queue nearest the gate!

During the week itself when the logs were extracted, I was in Ethiopia and Eritrea on behalf of Tear Fund on a short advisory

visit, looking at just the kind of projects English Woodlands have supported through Tree Aid. These two countries, which have suffered so much from war and famine, are gradually rebuilding the pitifully small remnants of forest through food-for-work projects. In exchange for a daily provision of grain, salt, and cooking oil, villagers give their labour to dig anti-erosion ditches and bunds, plant trees, protect young growth from browsing by sheep, goats, and camels, and establish tree nurseries. It's a long process but over the 13 years I've been visiting these countries for Tear Fund several thousand hectares of waste land have turned from brown and desolate to green and wooded, and once again become productive. Tear Fund's role, like Tree Aid's, is to help the local partner, in this case the Kale Heywet Church, to realise the vision and aspirations local people have in seeing their surroundings take a turn for the better. For the farmers and villagers themselves the planting and hard labour has created a wood of their own. Like ours, what they extract—fodder, sticks and poles, or grass for thatching—is theirs, though in their dry land the gathering of woodland produce will be a much less muddy business!

The long overdue thinning in Nain's copse has been done. The many remaining oaks have more room to develop and the promising ash and sycamore an opportunity to join them in a mixed canopy. Beneath the now dappled shade, splashes of light penetrate to the woodland floor and should help the patches of bluebells and primroses, always providing the roe deer don't find them first.

The oak in Nain's copse are thinned, but the freehold responsibility over all 30 acres of woodland continues. The cycle of forestry work follows the seasons though with each passing year there are tiny differences as the trees ever so gradually mature. Change is almost imperceptible until one looks at old photographs or someone visits who knew the wood years before. Both Alistair and John Newcomb have commented in just this way recently. As well as caring for and cultivating a forest crop, we have greater liberty to develop the wood to share its enjoyment more widely. All four of us who have an interest in at least part of it perhaps should plan how we might do this. There is no open water, so excavating a pond to encourage wildfowl, water lilies, and a little wetland would add wilderness and even woo woodland bats. It will probably occupy the next ten years, but the last ten have more than shown that owning a wood is far more than the sum of its timber.

16

Not only for timber

An unexpected delight of the last ten years has been sharing the wood with others for picnics, for camping, for wide games, for occasional church events, and even for charity fund raising. Sharing a woodland is quite unlike hospitality in the home or giving someone a lift in the car. It's also rather different from most farms. Friends visiting the wood can go anywhere at any time without upsetting the owners, though we would ask the impossible of those with dogs: to try and keep them under control and not to disturb game, especially in the autumn. There is no place that is out of bounds and no crops to trample, or livestock to frighten or be frightened by. For my wife, Margaret, this sharing has been the best thing about our wood:

> the principal enjoyment I have derived from the wood over the years is the pleasure we have given others by taking them there . . . Simple though picnic fare might be, the obvious enjoyment of families happily sharing a few hours in a peaceful environment, gives me great satisfaction.

The question most often asked when going over to the wood with friends or when someone finds out we own one, is 'What do you do with it?' The predictable exception are foresters who ask instead, in rapid succession, 'What are the tree species? How old are they? Is it profitable?' All these questions betray an interest in the unusual idea of owning a wood, on the part of many like ourselves brought up in suburbia with little day-to-day experience of rural crafts and country life. Picnics with friends have allowed many a leisured answer to these questions.

The first picnic spot was at the southern end of the cross ride in a grassy glade. Snatches of view through the gappy hedge gave glimpses over Litchfield tunnel to Popham airfield and the Little Chef on the A303 in the distance. The field on the other side of the hedge, and between where we sat and the railway, was usually down to cereals or, occasionally, the ethereal blue of linseed. Now it is the ground where Mike grows organic vegetables. Fifty-foot-high pines sheltered the glade, cocooning warmth and the aroma of resin on a summer's day. The site was a little suntrap. When it was really hot the scales of the pine cones opened crackling like far-off, sporadic small arms fire. In late spring, and usually the first time we would venture out for a picnic, the large hawthorn beside the endmost pine would be bedecked in may. In high summer interest would switch to the rose hips on the wild briar. Rabbits clipped the grass to a turf. On several occasions we also camped in this spot and learned that the trimmed lawn was far from even ground as the same rabbits had begun and abandoned numerous burrows.

As the hedge grew up, and as Mike used the adjoining field more intensively, the charm and the immediate view began to change. Also the ground was rutted when the pines were felled and dragged away, making the whole place less congenial. After trying one or two other locations we settled on a new picnic site in an equally sheltered spot at the bottom of the wood beside the railway. Until Taid's wood became ours this spot had not been available. This new site also enjoyed a backcloth of hawthorn with further shelter provided by clumps of overgrown hazel and the rapidly growing trees we had planted in 1987. The railway was more of an intrusion and definitely of much more interest to any youngsters bored by adult conversation or tired of climbing trees. Microlights from Popham and other aerial intruders continued to buzz lazily overhead, good flying conditions for them coinciding with good picnicking weather for us.

As I look back over the notes kept about the wood—not really a diary but jottings after each visit—it is possible to analyse scientifically that of the 421 visitors in ten years, 13 per cent have been relatives, 24 per cent church friends, 23 per cent foresters including scientists from my research station, and the rest, about 40 per cent, an assortment of acquaintances. Included in this quite pointless analysis are a TV presenter, a writer, a foremost exponent of Chinese brush painting, two families of missionaries going to Ethiopia and Mali, Sunday School teachers, a Cambridge professor, a mother of ten, and Britain's individually most successful sponsored walker according to the 1981 *Guinness Book of Records.* (Incidentally, the same edition has my photograph of a eucalypt tree native to Papua New Guinea which grew a staggering 100 feet in just $5\frac{3}{4}$ years—a world record at the time.) It is a pity we never kept a woodland visitors book! The picnics have been entirely spontaneous affairs and not modern soirées to accumulate personalia. But the value of keeping a record has shown itself as every true diarist knows and one American president found to his impeachable cost.

Camping in the wood is altogether different. The absence of services of any kind deny spontaneity and sudden impulse. Water, in particular, has to be brought with us and then husbanded carefully between drinking, washing, and washing up though the boys' interest in it rapidly diminishes in pretty much this order. We normally camp for just two nights and even then a 3-mile walk to the Popham Little Chef solved the main meal on day two. Excepting this one indulgence we were real campers. No camping gas, no large cubical tent, no portaloo, and no car with trailer, just a Force 10 tent with flysheet and the challenge of lighting a wood fire with a single match.

Most camps have been with Stephen and Ben and their friends. Jon was well immersed in such outdoor skills from his time in the scouts, but our two younger boys had not wanted to join. Both Margaret and I had been enthusiastic in guiding and scouting; indeed, we first met at opposite ends of a ping pong table on a joint scout/guide do in Petts Wood. It was up to us to introduce backwoods skills to our younger boys. And, of course, it has allowed us to rediscover and pass on the youthful pleasure of a properly erected tent, how to get comfortable when only a groundsheet and sleeping bag separates body from mother earth, and showing how to light a fire with one match. This last was very much Dad's work, and in the wood a particular challenge without any holly bushes to raid for the dry kindling always to be found in their interior. A good second best is to peel tissue-thin slivers from the bark of young birch. It's rather like a natural paper, but it curls up and is so

light that it needs quickly covering with the finest dry twigs available, with coarser ones on top. Although there are a few ash in the wood big enough to provide the best of all firewood even when freshly cut, dead beech from the attentions of rabbits, squirrels, chlorosis, and beech bark disease always provided plenty of dry wood. We never needed to resort to pine branches which, although igniting easily, spit sparks and burn too rapidly. The carefully laid fires almost always caught alight with the one-match ration. We adopted too the scouting ploy of greasing the bottoms of the cooking billies with washing up liquid so that the blackened pots dissolved to silver at the final washing up without needing Brillo pads. Despite these little economies Margaret didn't join us on these expeditions but was attentive in provisioning our every need, including the suggestion to go to the Little Chef. She knew that even the delicious essential of all camps, a breakfast fry-up of bacon and eggs, would not keep us going all day.

Camping in the wood usually led to a poor night's sleep. Eyes and ears were alert to every movement and strange sound; the ground got harder as the night wore on, and the unfamiliar confinement of the sleeping bag made turning difficult for my middle-aged shape. And that was not all. At about 5 a.m. goods trains would clatter past, and just over an hour later the up-train from Winchester would inaugurate the new day's commuting to London. Although the campsite is about 100 yards from the railway, night-time amplification made the roar entering Litchfield tunnel sound like a roar entering our tent door. What the brain knows and what the ear hears seemed a world apart. Half-waking, dreamily the train was jumping from its rails and crashing towards the tent. Wide awake and it was gone, no more than a receding echo. The day had begun early, at least for me, the boys having stayed up late were still dreaming and would need more than a distant train to wake them. I couldn't even go downstairs to make a cup of tea.

On three occasions, while Jon was a scout, the hardier members of the 3rd Alton troop used the wood for one night winter camping experience any time between January and early March. The lads were always welcome and would do a bit of branch pruning or cut back a few overgrown hazel by way of thanks. I was also grateful to their leader, Sandy, for lending me the crowbar used to open up the holes for the treeshelter stakes when planting Taid's wood. At two of their camps several scouts were taught elementary forestry to help towards a badge. I don't know whether it has stimulated any of them to take up forestry as a career as my own experience did 30 years before.

In June 1988, which was a dry warm month, one of the largest gatherings in the wood took place. My wife's father, Robert Steel, was shortly to embark on another of his long walks in Britain to raise funds for charity. He had done so well at this in 1979 that he made the 1981 *Guinness Book of Records,* having raised through sponsorship £71 210, walking 1001 miles from John-o'-Groats to Land's End for the benevolent fund of his own profession, the Royal Institute of Chartered Surveyors. He has also walked the entire perimeter of England to raise funds for the National Trust. This time the fund raising was in support of the Lord Mayor's appeal to help Great Ormond Street Children's Hospital, Action Research for the Crippled Child, and the Lord Mayor Treloar College for disabled children. We invited many friends for a picnic and Bob briefed us about his plan to walk exactly 1200 miles down the spine of Britain from Strathy point on the north coast of Scotland to Portland Bill in Dorset, and then abruptly turn north-east on to London's Mansion House, the residence of the Lord Mayor. The whole route traced an inverted figure 7.

My father-in-law is still planning long walks, even at the age of 75. In 1993 he was refused permission to join the walkers celebrating the opening of the channel tunnel because he was over 65, even though all the above walks were done after this age! To make up for this he is undertaking his most ambitious walk yet: a perambulation around the perimeter of Great Britain—The Orbita Britannica Walk—a total distance of 4444 miles, to mark the centenary of the National Trust and raise funds for their coast and countryside programmes.

The most exacting visit to the wood was when colleagues of my own profession, members of the Wessex Silvicultural group of the Institute of Chartered Foresters, included it in one of their study meetings. Such meetings occur several times a year, and by various groups in all parts of Britain, to help foresters learn their trade by seeing the handiwork of fellow foresters. Woodlands are looked at, species choice considered, thinning practice commented on, provision for wildlife noted, and the latest information on grants mulled over. The hapless owner presents data about each compartment of the wood, or on the particular matter under study such as wildlife conservation, or timber harvesting, or a chosen tree species. The typical sized party of 20 or 30 pass comments, opinions and observations which are, for the most part, constructive. Nevertheless, it is like having one's work of art or a composition publicly examined and dissected. In our case the Wessex group were generous, commenting on the need to thin the rather dense pine and beech mixture—it was just before the pine were cut out—and expressing delight at the excellent growth of the new planting in Taid's wood. Visits with groups such as the Wessex have taught me much practical forestry and I owe a great debt to the many owners and foresters whose woodlands I have inspected and picked over: it has been a remarkable living textbook to add to the many of the more conventional kind I have worked through in my student days and research career.

As well as the Wessex group, members of the Irish forestry service 'Coillte' have inspected our efforts. Colleagues from the Alice Holt research station have on three occasions come for afternoons of games and refreshment, but each time conversation turns to discussion on points of tree pathology or woodland management or further quizzing on the wood's profitability. From time to time, our wood has been examined for particular silvicultural questions, such as the problem with plastic treeshelters not breaking down quickly enough. My solution of using a Stanley knife held tangentially to slit the tubes ended up as a recommended method illustrated in Gary Kerr's article in the trade journal *Forestry and British timber*. A photograph of the cleanly prepared site ready for planting Taid's wood features in Farming Press's book *Farm woodland management*.

Most visitors arrive by road, but one did come by air and leave an unusual calling card which we surprised upon during one picnic. It was the tag from a balloon which a hopeful youngster at Mrs Blands County Infant School, Burghfield Common near Reading had launched skywards at a spring fête. It must have travelled about 30 miles to reach the wood driven by a northeasterly breeze. By the time we found the card, mice had already done so as well and chewed part of the reply section. Enough was still left to identify the white balloon as number 132 and one day was left before closure of the competition. The card was promptly sent

off by first class post. Five weeks later we learned that balloon number 132 had travelled the furthest and we received a gift voucher: I hope the successful youngster did too!

Three visitors to the wood have been a special help. My mother, my niece Barbara who is a botanist, and Rod Stern, a forestry colleague and expert on mosses, have all prepared lists of flowers and other plants to be seen. Adding these together—the lists were made at different times of the year—along with my own observations, the wood sports at least 75 different flowering plants and 46 mosses, lichens, and liverworts, but only one of which is really unusual. This is not unexpected since the woodland is not ancient and the rich diversity of flowers are those common to newly colonised ground in chalky and limestone country. The one rarity is the tiny moss mentioned in Chapter 14 and not known anywhere else in Hampshire. The moss itself is an introduction from America first seen in Britain in the 1960s, and it has now turned up in Hampshire. In general the lichen flora of the wood is disappointing, probably owing to the pervasive emissions from oil refineries and other industry some 30 miles upwind to the south-west. Lichens are good assessors of air pollution and our part of Hampshire is apparently influenced by whatever it is that issues forth as an unseen, unwelcome aerial visitor.

My mother takes a special delight in the wood and visits it as often as she is able to. Countryside has been a joy and long walks in Kent and Surrey always held for her that harbinger of some new experience or refreshment of soul just round the next corner. As a boy I remember my mother getting up at dawn in June and July and enjoying the chorus of birds which came into the garden. She would rise that early in summer to mark O-level exam scripts before the family's day began. My stamp collection benefitted from the heavy parcels that had been arriving in the weeks preceding. On Saturdays we would sometimes take a picnic to the National Trust's Petts Wood, the entrance to which was at the bottom of our road, and mother would sit and mark papers and we would climb trees or simply muck about. The pain at so much loss in the great storm of 1987, in this particular and much loved woodland near where she lived for half a century, was set down in verse and is in Chapter 9. She is of a generation properly taught to learn by heart and to recite substantial passages of prose or whole poems. She has committed to memory more than 2000 lines of verse, though I've not heard her at one sitting. For our younger two boys, Stephen and Benjy, she composed, as she called it, this jingle:

Down in your very own woodland
  And in sunlit glade—
The leaves on the trees were too young
  To cast any shade—
We wandered through springtime glories
  In one of life's golden hours,
And when we came back from your woodland
  My pockets were full of flowers.

You had dropped them so gently in,
  (Benjy and Stephen too)
The tiniest sprigs of flowers,
  Yellow and pink and blue.

At the end of a golden hour
  We left that lovely spot,
Left its flowers to themselves:
  Ground ivy, forget-me-not,
Primrose, violet, daisy,
  Dandelion, groundsel too,
In patches and clusters and cushions,
  Pink and yellow and blue,
And, best of all, on the very edge
  Of what we call Taid's wood,
Among a scatter of twigs and leaves
  Three cowslips stood!

Often in dark, grey hours
  I shall think of that golden day
When we stood in a primrose glade,
  And a Brimstone came our way.

Joyce F. Evans
23 April 1988

My mother's enjoyment of the wood did not end with her visits or even this engaging, poetical appreciation, as she has helped in practical ways with her inventories of flowers, sightings of butterflies and, of course, the new copse we have named after her.

A wood is not only for timber, as we have certainly experienced. Woodland fruits of hazel nuts, apples from the one tree in the very middle of the wood, and, in 1994, massive quantities of blackberries have been forthcoming. For several years we kept relatives supplied with Christmas trees from tops of the Douglas firs, but have now run out and I've not got round to planting any. Bean sticks are cut from the overgrown hazel. And firewood warms several homes, helping to mitigate the rising levels of carbon dioxide by reducing in the minutest amount the extravagant consumption of fossil fuels. Wood, unlike coal, oil, and gas, is renewable.

But it is the sharing of the wood with many people, and now with you, that is best of all. Our prayer has been answered. Really, the privilege is that we've been allowed a share in the life of the wood. It was here long before we were and, God willing, it will be for a long time after us. In that sense, it's not really a wood of our own at all.